D0828322

VoIP

VoIP

Wireless, P2P and New Enterprise Voice over IP

Samrat Ganguly
NEC Laboratories America Inc., USA

Sudeept Bhatnagar
AirTight Networks, USA

John Wiley & Sons, Ltd

Copyright © 2008   John Wiley & Sons Ltd, The Atrium, Southern Gate, Chichester,
West Sussex PO19 8SQ, England

Telephone   (+44) 1243 779777

Email (for orders and customer service enquiries): cs-books@wiley.co.uk
Visit our Home Page on www.wiley.com

All Rights Reserved. No part of this publication may be reproduced, stored in a retrieval system or transmitted in
any form or by any means, electronic, mechanical, photocopying, recording, scanning or otherwise, except under the
terms of the Copyright, Designs and Patents Act 1988 or under the terms of a licence issued by the Copyright
Licensing Agency Ltd, 90 Tottenham Court Road, London W1T 4LP, UK, without the permission in writing of the
Publisher. Requests to the Publisher should be addressed to the Permissions Department, John Wiley & Sons Ltd,
The Atrium, Southern Gate, Chichester, West Sussex PO19 8SQ, England, or emailed to permreq@wiley.co.uk, or
faxed to (+44) 1243 770620.

Designations used by companies to distinguish their products are often claimed as trademarks. All brand names and
product names used in this book are trade names, service marks, trademarks or registered trademarks of their
respective owners. The Publisher is not associated with any product or vendor mentioned in this book. All
trademarks referred to in the text of this publication are the property of their respective owners.

This publication is designed to provide accurate and authoritative information in regard to the subject matter
covered. It is sold on the understanding that the Publisher is not engaged in rendering professional services. If
professional advice or other expert assistance is required, the services of a competent professional should be sought.

Other Wiley Editorial Offices

John Wiley & Sons Inc., 111 River Street, Hoboken, NJ 07030, USA

Jossey-Bass, 989 Market Street, San Francisco, CA 94103-1741, USA

Wiley-VCH Verlag GmbH, Boschstr. 12, D-69469 Weinheim, Germany

John Wiley & Sons Australia Ltd, 42 McDougall Street, Milton, Queensland 4064, Australia

John Wiley & Sons (Asia) Pte Ltd, 2 Clementi Loop #02-01, Jin Xing Distripark, Singapore 129809

John Wiley & Sons Canada Ltd, 6045 Freemont Blvd, Mississauga, ONT, L5R 4J3, Canada

Wiley also publishes its books in a variety of electronic formats. Some content that appears in print may not be
available in electronic books.

Library of Congress Cataloging-in-Publication Data

Ganguly, Samrat.
    VOIP: wireless, P2P and new enterprise voice over IP / Samrat Ganguly, Sudeept Bhatnagar.
        p. cm.
    Includes index.
    ISBN 978-0-470-31956-7 (cloth)
1. Internet telephony. I. Bhatnagar, Sudeept. II. Title.
    TK5105.8865.G36 2008
    004.69'5–dc22

                                                                                                            2008007367

British Library Cataloguing in Publication Data

A catalogue record for this book is available from the British Library

ISBN 978-0-470-31956-7 (H/B)

Typeset by Sunrise Setting Ltd.
Printed and bound in Great Britain by Antony Rowe Ltd, Chippenham, England.
This book is printed on acid-free paper.

CONTENTS

PART IV   VOIP IN ENTERPRISE NETWORKS

PART V   VOIP SERVICE DEPLOYMENT

16 Supporting Services and Applications   223

17 Security and Privacy   231

PREFACE

Voice over Internet Protocol (VoIP) is rapidly becoming the technology of choice for voice communication. Several books cover the topics of specific components of VoIP in detail. At a basic level, most of the books in this space describe *how* VoIP and its various components function.

However, we feel that there is a huge void with respect to information that helps a reader to understand *why* certain features are present, *what* is the quantitative impact of the existing design choices, and *how* the next generation VoIP should evolve. Knowing how certain components work merely gives a partial view of VoIP and not a complete well-connected picture to the reader to get a whole perspective. In this book, we try our best to bridge this gap in the VoIP information space.

Focus of the book

We stress that this book is about *concepts* that underly VoIP and its components. It is about the *performance* of VoIP components in different real-world settings. It is about understanding the *real-world facts and constraints* that should guide the design choices in a VoIP deployment. We highlight the performance issues faced by VoIP owing to the underlying network technologies. We make extensive use of experimental results from recent research showing the impact of various technologies on VoIP performance. The book is *not* meant to describe the specifics of all protocols and components used in VoIP and therefore, is *not* a comprehensive reference manual covering intricate details of any technology.

The book is written in such a manner as to focus on the concepts when describing the components and their interactions, and subsequently highlight the actual system performance under different design choices. Where necessary, we give a brief overview of the specifics of protocols and standards, and will give adequate references at the end of each chapter to guide a user interested in knowing more about the specifics of the topics discussed in the chapter. The book is meant as a guide that provides insights into a wide range of VoIP technologies for a reader intending to understand the technology.

Intended audience

This book is our attempt to disseminate the information that will help the reader to gain a deep understanding of VoIP technology. The content of this book will help an engineer deploying the VoIP technology to acquire substantial knowledge and be able to make informed design choices. This book will help a student who aims to become a VoIP system designer rather than a system deployment technician. A technical reader who is not interested in the nitty-gritty details and needs to have a big picture of the VoIP arena will gain immensely from this book.

Guide to the chapters

The book is organized into five logical sections that describe the impact of diverse technologies on VoIP. The first section introduces the basic components of a VoIP deployment. The next two sections focus on the underlying IP networks (Overlay and Wireless Networks). This focus is intended to provide a general awareness of what each network provides in terms of supporting VoIP. At the same time, the two sections provide a deep understanding of how the network level characteristics affect VoIP and how various network-specific deployment issues are being addressed. The following section on VoIP in Enterprise Networks covers the aspects that are relevant mostly in the VoIP deployment in enterprises. The last section details the relevant auxiliary issues related to the deployment of VoIP as a service.

Each section begins with a summary page that defines the scope and organization of the chapters and introduces the chapter content. Each chapter starts with a brief introduction to help the reader to get a feel of what to expect and ends with a summary to provide a set of simple 'take away messages'.

S. GANGULY, S. BHATNAGAR

Princeton, NJ

PART I

PRELIMINARIES

This section deals with the basics of VoIP technology. The chapters in this section give an overview of the various issues and technologies underlying VoIP, ranging from the fundamentals of the Internet to an overview of various aspects of VoIP.

While each chapter in this section contains several topics that can be elaborated significantly more, we refrained from doing so. The goal of this section is to provide a background to the reader and have a framework in which we can place the rest of the book.

The section starts with Chapter 1 giving an overview of the fundamental concepts that are required for any telephony network with a reference to the legacy telephone network. The chapter further describes the fundamentals of the current Internet and shows how it can be utilized as a telephony network. Chapter 2 gives an overview of the working of VoIP using a generic architecture. It further provides a glimpse of various issues that any VoIP deployment must tackle.

We delve into the details of voice codecs in Chapter 3. This chapter shows how the analog voice signal is converted into digitized packets for transportation over the Internet. A range of codecs with different capabilities and limitations are highlighted. The actual performance of these codecs in diverse conditions using extensive experimental results is discussed in Chapter 4.

CHAPTER 1

INTRODUCTION TO VoIP NETWORKS

Voice over Internet Protocol (VoIP) has exploded onto the technology scene in the past few years. VoIP is set as the technology that takes our current telephony system referred to as Public Switched Telephone Network (PSTN) to the next generation. Before delving into how VoIP stands to deliver on that promise, we take a brief look at telephony in the PSTN space. Our discussion of PSTN will be more conceptual rather than merely elaborating the components and protocols. The goal is to make the reader understand the philosophy that drove the design of the telephony network and also to lay a foundation to the type of services that would be expected of a full-fledged VoIP network.

1.1 PUBLIC SWITCHED TELEPHONE NETWORK (PSTN)

The era of telephone communication started in 1876, when Alexander Graham Bell enabled the transmission of voice over a wire connecting two phones. Fundamentally, the role of a telephone connection in completing a call is very simple – it needs to connect the microphone of the caller to the hearing piece of the receiver and *vice versa*. In the beginning of the telephony era, each pair of phones had to have a wire between them in order for them to communicate. There was no shared component between the devices, so while people wanted telephones to communicate, the system was not cost-effective.

VoIP: Wireless, P2P and New Enterprise Voice over IP   Samrat Ganguly and Sudeept Bhatnagar
© 2008 John Wiley & Sons, Ltd

1.1.1   Switching

Perhaps the most important development that proved to be a huge step in making a large-scale telephone system viable was the concept of a *switch*. The insight that drove the design of a switch was that a dedicated wire between two telephones was essentially being used for a very small fraction of time (unless the parties at the two ends talked all day on the phone); so a way of using that line to serve some other connection while it was idle would serve to reduce the cost of deployment. In particular, the concept of *multiplexing* was used. The idea was to be able to share the line between multiple telephones on an on-demand basis. Of course, the trade-off was that if two pairs of phones were sharing a single line, only one pair of them could talk at a time. On the other hand, if most of the time only one pair of them intended to communicate, the telephone system could do with only one line rather than two.

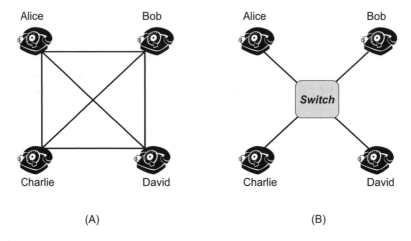

Figure 1.1   Basic functionality of a switch over a four-telephone network. (A) Without a switch six lines are required to connect the four telephones. (B) With a switch, only four lines are required to connect the four telephones to the switch.

In order to implement the concept of multiplexing, there was another problem to be solved. While sharing of a line was definitely a nice insight, the whole wire could not be shared end-to-end since its endpoints are two of the telephones residing at diverse locations. This led to the concept of segmenting the end-to-end wire into smaller pieces and applying the sharing logic onto these pieces. The device that connected these pieces together is a switch. Consider the example shown in Figure 1.1. There are four telephones that need to communicate with each other. In Figure 1.1(A), they are connected directly to each other requiring a total of six lines. In order to reduce the number of lines required, the lines are broken into smaller segments and connected to a switch. as shown in Figure 1.1(B). Now no two phones are directly connected to each other. When Alice wants to talk to Bob, it is now the switch's responsibility to connect the corresponding two segments together so that together they act as an end-to-end wire. Note that if Alice and Bob are talking, and Charlie wants to talk to Bob, then he cannot at that instant. This is because the line segment from the switch to Bob is already in use for the call from Alice. Thus, Bob's phone is 'busy'.

In the early days, the function of switching was done manually. There was an operator who connected the lines together to provide connectivity. As technology advanced, these switches were automated and were able to switch several calls simultaneously. The switches today are electronic and very adept at their task while handling hundreds of calls simultaneously.

1.1.2   Routing

While the concept of switching was an important driver in making the telephony viable in a small geographic region, it was still not enough to spread to larger areas. This is because it was not feasible to connect all the phones in the whole area (state, country, world) to a single switch. This implies the need to have multiple switches corresponding to diverse geographic regions. Of course, this also means that if Alice's phone is connected to switch A and Bob's to switch B, then for Alice to call Bob, both switch A and B have to connect their respective segments (to Alice and Bob respectively) as well as to have a connecting segment between them. Conceptually, this requires that each pair of switches should now have a link between them to allow all pairs of telephones to be able to communicate with each other.

Again, the requirement for all switches to connect to all other switches is not scalable. For example, it may be reasonable to have a line connecting switches to two neighboring cities. However, having a line from each switch to every other switch in the world is infeasible.

The alternate strategy extends the concept described earlier. When two phones connected to two physically connected switches need to talk, we required three line segments to be connected together: Alice's phone to switch A, switch A to switch B, and switch B to Bob's phone. However, if switch A and switch B are not directly connected, they can still be able to connect through a chain of switches in between. Thus, the larger the distance between Alice and Bob, the larger the number of switches in the path between them. Conceptually, when Alice calls Bob, a whole set of segments and switches are connected in sequence to provide the feel of an end-to-end wire between the two of them. None of these segments can be used for any call while this call is in progress. Essentially, this results in building a dedicated *circuit* between Alice and Bob.

The above example uses links which can carry a single call at a time. In practice, the switch-to-switch links (also referred to as exchange-to-exchange links) are replaced by *Trunks* that can carry multiple calls simultaneously. This is achieved by methods such as Time Division Multiplexing (TDM) where frames from different calls (containing the encoded voice signals) are multiplexed into a TDM frame that runs over a higher bandwidth. This results in the perception of all the calls proceeding simultaneously. Although the number of calls carried in a trunk is much higher, the bandwidth limitations of the medium limit the number of simultaneous calls possible over a trunk as well. For example, in the USA, a TDM frame contains 24 voice frames implying at most 24 simultaneous calls over the corresponding trunk.

1.1.3   Connection hierarchy

With the help of call routing, any two telephones can communicate over a sequence of switches. However, how do we decide which switches are to be connected to each other? Consider a simple case, where there are three switches A, B, C. Physically, A and B are closer to each other (say in adjacent cities) and C is far away from both of them (another country). One possible connection could be to have $A \leftrightarrow C$ and $B \leftrightarrow C$ as the two links.

Now any call from a phone connected to A to a phone connected to B will have to be switched across the country to C from where it would be routed back to B. Clearly this type of long link should be avoided as much as possible.

This implies that the switches within each other's vicinity should be connected to each other rather than to those far apart. A natural extension of this philosophy implies that the switches within cities should be connected to form a network, a few access nodes from this network should connect to other networks in similar states, and the same philosophy extends to countries. Essentially, the political boundaries themselves serve as guidelines to forming networks of switches.

1.1.4 Telephone numbering

Once the hierarchical organization of switches and, in general, exchanges is decided, the final piece of the puzzle is to figure out where a particular phone is located in order to call and how the corresponding call should be routed. Across a large network spread across the globe, knowing all the routes to all the other switches and destinations is not feasible. Thus, each switch can know only a few neighboring switches.

The problem of routing in such a scenario is automatically solved using a proper telephone-numbering system (E.164) that we use today. For example, a telephone number consists of a code, an in-country zone code, and a number describing the local switch/exchange to which the phone is connected. Using the digits of the phone number, the switch at the caller's end would know to which of the neighboring switches the call should be routed. Following the same procedure end-to-end, a VoIP call is easily established.

1.1.5 Signaling

The call setup procedure described above requires some means of informing all the devices on the end-to-end path of the call to switch the call accordingly. This is achieved using signaling. The current telephony network is based on sophisticated signaling protocol called SS7. It is the most prominent set of protocols in use in the PSTN across the world. Its main use is in setting up and terminating telephone calls. SS7 uses an out-of-band signaling method to set up a call. The speech path of the call is separated from this signaling path to eliminate the chances of an end-user tampering with the setup protocol.

In the PSTN, telephones constantly exchange signals with various network components such as dial tone and dialing a number. SS7 facilitates this type of signaling in the current PSTN. In general, SS7 forms the core of the current PSTN. Along with call establishment and termination, it provides the aforementioned functionalities such as call routing.

1.1.6 Summary

Our description of traditional telephony describes the most important concepts required in setting up a voice call across the network. Switching allows the telephone to multiplex over a limited number of links. With switches connected to each other indirectly, routing is required to set up a call over multiple hops. Using the concept of hierarchy, the E.164 numbering assigns a logical method to discover users and to set up end-to-end phone calls. In order to set up any end-to-end call, the devices on the path of that call need to take appropriate action to switch the call correctly so that it follows the set route. This entire setup is attained using signaling.

It is important to see that while these concepts are described as applicable to PSTN telephony network, in fact, any large-scale telephony network needs to provide these functions. Thus, enabling VoIP over the Internet (which is a large-scale network) also implies that these functions be provided in the Internet. We shall look at how these functions are provided in the Internet both in general and specifically for VoIP.

1.2   FUNDAMENTALS OF INTERNET TECHNOLOGY

What we described above gives the basic idea regarding any telephony system. To enable voice over an IP network such as the Internet,[1] the capabilities described above need to be provided in the Internet as well. In the following, we describe how these functionalities are provided in the Internet in general.

1.2.1   Packetization and packet-switching

The PSTN is based on the concept of *circuit-switching*. For any call to go through, a complete end-to-end path is set up comprising of intermediate switches and the dedicated links between them. This sets up a path that is specifically meant for the call prior to the user being able to communicate. Having such a dedicated *circuit* for a call means that the delay faced by each signal element (carrying the voice) is constant. The components used in the circuit are not available for use until the call terminates.

Each circuit has a capacity to carry some amount of information at each instant. In case of voice, this information is the signal containing the encoded speech. Dedicated circuit for a call results in a wastage of the capacity even if for a small time it is not being used to carry information. This wastage is more prominent in case there are other calls that are not able to connect for want of a segment of this underutilized circuit.

In order to overcome this capacity underutilization a new switching method was conceived. The switching method is called *packet-switching*. The idea is to *packetize* the information, i.e. break down the information to be transmitted into smaller chunks called packets and send each packet independently towards the destination. There is no dedicated end-to-end path setup prior to the communication. Each packet containing information to be delivered is sent towards the destination. Each switching element *router in the Internet world*, would look up the destination address in the packet and send it to the next switching element on the path to the destination. Essentially, different packets belonging to the same end-to-end communication session can take different paths in the network, since there is no circuit for them to follow.

The efficiency gain from circuit switching are from multiplexing at a fine level. Since there are no resources reserved for any end-to-end session at an intermediate router, the router treats all arriving packets as equals. The packets from different sessions are lined up in a queue inside the router which decides where to send the packets one by one. Thus, the router is being used by all calls simultaneously. In the case of VoIP, think of packets containing voice from two different calls sharing the router queue. Also, packet-switching does not lock the router (and a link) for a particular call, implying that packets from a second call can be switched if there are no packets from the first call using the router.

---

[1]While the Internet is an embodiment of an IP network, we shall use the two terms interchangeably throughout the book.

Packet-switching forms the backbone of the Internet. The computers (end-hosts) from the end points of communication are connected using routers. Two computers communicate with each other by packetizing the information to be sent out and then send each packet to the network. They are routed by the network to the destination without establishing an end-to-end connection *a priori*.

1.2.2 Addressing

In its present form, the most prevalent addressing scheme in the Internet is based on Internet Protocol version 4 (IPv4). The allocated addresses are called the IP addresses of the respective devices. Each IPv4 address is 32 bits (4 bytes) long. Each address can be written in the dotted decimal notation as A.B.C.D where each of A,B,C,D is a number between 0 and 255 (representing 8 bits).

The addresses are allocated to organizations in sets defined by the common prefix shared by the addresses in the set. In the initial era, the addresses were categorized into classes and allocations were at the granularity of classes. The classes to be allocated for unicast addresses were called Classes A, B and C. Class D defined multicast addresses and Class E addresses were reserved for future use. Each Class A group of addresses was identified by its first 8-bit prefix and hence contained 2^{24} distinct addresses. Similarly Classes B and C had 16- and 24-bit prefixes resulting in address spaces of 2^{16} and 2^8, respectively. Assigning address spaces at this granularity had an adverse impact as the available address space started depleting very quickly.

To address this concern, a new proposal called Classless Inter-Domain Routing (CIDR) was introduced where the IP addresses were allocated in chunks and identified by their prefixes rather than classes. Thus, a large chunk of addresses that contain all addresses starting with the first 9 bits being 100001011 is written as 133.128.0.0/23 where the 23 represents the number of bits that can vary with the prefix 9 bits fixed to 100001011 (133.128 represents the decimal value of 1 000 010 110 000 000).

In CIDR addressing, a large chunk of addresses is now given to allocation authorities that can create smaller chunks out of it to allocate to the organizations. For example, a country can be allocated a chunk 133.0.0.0/24. From this chunk, organizations can be given smaller chunks which may depend on their location in the country. This will automatically construct a hierarchy of addresses. The major benefit of such a hierarchy is seen in the routing efficiency.

1.2.3 Routing and forwarding

Routing refers to the process of computing the routes between any two hosts. In a router, the routing process fills out a *routing table* (or forwarding table), that contains information about which interface the router should forward a packet to so that the packet reaches closer to its destination.

There are two types of routing protocol in the Internet: intra-domain and inter-domain. The intra-domain routing protocols (such as RIP, OSPF) operate in a single domain under the control of one administration. In a domain, the messages contain information about the connectivity information of all the nodes in the domain. After applying the correct routing algorithm such as Dijkstra's shortest path algorithm, the routes are computed and the routing table of each router is populated. Inter-domain protocols (such as BGP) operate on a coarser granularity. The border router of a domain provides a list of prefixes to which it can route. Based on the routing policy, the routers will decide the routes for these prefixes. An interesting observation here is that there is implicit hierarchy here. Inside a domain each router knows every other router and would also know the IP addresses of all hosts that are directly connected to this network. However, the external network appears as a single entity in the form of a prefix advertised by a neighbor.

The routing table of each router is computed using the intra-domain and inter-domain routing protocol. Since the Internet is a packet-switched network, the goal is to be able to route any packet to its destination. The routing table is the core that allows this. It contains information about where a router should send a packet, based on its destination IP address. An interesting thing to note is that if the routing table contains an individual entry for each destination IP address in the Internet, there will be 2^{32} entries in the routing table. It is very difficult to manage this number of entries in a router. The CIDR-based scheme allows routing tables to be compacted. In this case, the adjacent prefixes could merge into single entry if the corresponding outgoing interfaces for both sets of prefixes is the same. The exact packet header matching algorithm is called *Longest Prefix Match*. If there are multiple entries in the routing table that match the destination address of a packet, then the entry which has the maximum number of prefix bits common with the destination IP address is considered the valid matching entry and the packet is forwarded accordingly. For example, for a packet with destination address 133.193.20.24, if there are two entries in the routing table 133.192.0.0/24 and 133.193.0.0/16 (with corresponding forwarding interface), both will match with the packet's destination address. However, we will use the entry with the latter prefix as it has more prefix bits in common with the destination IP address and forward the packet to the interface corresponding to that entry.

When a packet arrives at a router, the following functions are performed in order:

- *Routing Lookup:* At the incoming interface, the router needs to determine the output interface for the packet. The router uses the longest prefix match to find the most specific entry in the routing table corresponding to the packet's destination. A lookup on that entry gives the output interface to which to send the packet. Using the router's switching fabric, this packet is sent to the corresponding output interface.

- *Queue Management:* Each output interface has a buffer where it queues all packets forwarded to it by all incoming interfaces. The buffering is required because the output link capacity might not be sufficient to handle the combined traffic from all interfaces. Since the buffer size is finite, the buffer could be full when a new packet arrives. The basic task of the queue management strategy is to determine *which* packet to drop in such a case. Traditionally, the routers follow a *drop-tail* policy where the incoming packet is dropped if the buffer is full. Note that this is also an implication of packet-switching as the buffer is being shared by packets from diverse connections. Of course, if the buffer is not full, typically the packet will be enqueued at the back of the queue of packets that already reside in the buffer. This is not always the case because there are certain *Active Queue Management* mechanisms (such as Random Early Detect – RED) where an incoming packet may be dropped even if the buffer is not full. This is done to indicate to the host that congestion is imminent and it should slow down its traffic rate.

- *Scheduling:* The scheduler resides on the output interface of a router. Its task is to select a packet from the queue to transmit. In the current IP routers, the predominant scheduling policy is *First In First Out (FIFO)*. Thus, the scheduler picks the first packet in the queue and sends it out on the link. However, from the perspective of VoIP, it may be beneficial to send voice packets, which may be at the back of the queue, prior to the other non-real-time packets such as those belonging to an FTP session.

1.2.4 DNS

Domain Name System (DNS) provides the name to which address the mapping service in the Internet. It is one of the most important services in the Internet. DNS provides the service equivalent of directory lookup.

A DNS query takes a Fully Qualified Domain Name (FQDN) such as a URL and the response contains the current IP address associated with the given FQDN. One of the most important benefits of DNS is that it allows users to remember easy-to-memorize strings rather than IP addresses. For example, it is much easier to remember the website for Wiley as `www.wiley.com` rather than remembering a set of four numbers representing its IP address.

While from the perspective of the user this FQDN to IP address translation is the single advantage that DNS provides, it has several features from the perspective of the service providers. It allows the servers handling different services in an organization to be identified. We shall see more details of one such usage in Chapter 16. Furthermore, it allows load balancing across servers by returning different mirror server addresses to different user queries. This provides a simple load-balancing solution. As a further extension, the same applies to the use of DNS to provide fault tolerance. If one of the servers providing a service fails, DNS can be used to provide the address of another server seamlessly.

In order to provide this basic service, DNS essentially serves like a distributed database. The Internet namespace is divided into *zones* with the responsibility of managing the namespace in each zone being delegated to a particular authority. Thus, a zone is essentially a unit of delegation. For example, the authority of the `.com` zone is delegated to a single authority and that of the `wiley.com` zone is delegated to another authority. Each zone can have one or more DNS servers which maintain the local namespace database. For example, the name to IP address mapping information for `www.wiley.com` would rest with the DNS server for `wiley.com` zone.

A DNS request can originate from any host in the Internet. In the simplest case, the DNS query would process the text of the FQDN from right to left. Thus, to query for the IP address of `www.wiley.com`, a host would first go to the *root domain server*. That server will redirect it to one of the top-domain servers for the `.com` domain. The `.com` domain server would inform the user to query the `wiley.com` domain's server which will have the answer to the query. In practice, this process is augmented with caching. When a query is issued (say by the web browser visiting a web site), it calls a *resolver* software in the local machine. The resolver usually caches some popular FQDN-to-IP matches that some prior DNS lookup had returned. If the current query is satisfied by a cached entry, the resolver returns that address. If not, it forwards the query to the preconfigured DNS server that the host's ISP has provided. Again that DNS server maintains a cache of frequently resolved FQDN-to-DNS mappings. If the query is not answered from its cache, it follows the aforementioned procedure as a client would, and returns the result to the host.

1.3   PERFORMANCE ISSUES IN THE INTERNET

While the Internet provides all the features that are required of a telephony network, there are significant other problems that it introduces. It may be tempting to think that with switching, routing, addressing and lookups being provided, VoIP would have no special concerns in the Internet. However, this is not correct. In fact, while the cost of deploying VoIP over the Internet is considerably less (as it is using a shared network), there are significant performance issues that need to be addressed. For VoIP to be a viable alternative to the PSTN, not only should it be cheaper and easier to deploy and maintain, it should provide similar or better call quality so as to motivate an end-user to move to VoIP.

The performance issues that the Internet faces stem from its packet-switching nature. Packets from several flows share the queue at the output interface of a router. The bandwidth that the link connected to that interface is limited. Thus, the resources of the router are shared, resulting in several performance glitches.

1.3.1   Latency

Latency is the total delay that a packet faces while it travels from its source to its destination. There are multiple contributors to the latency. The foremost of these contributors is the physical limit imposed by the speed of light (or electromagnetic wave, depending on the carrier). For example, if a packet (or a signal) has to travel 3000 km over a link, then at speed of light (300 000 km/s), it will take 10 ms to travel. In practice, the signal travelling speed is lower than the speed of light. This delay has to be faced independently of the underlying signal-carrying technology. The second contributor to the latency is the *queueing delay*. This is the delay that a packet faces at a router when it is stuck behind other packets waiting for its turn to be transmitted. Note that this delay is not present in the circuit-switched network where there is a dedicated circuit present for the signals for a call. The last source of latency is the *transmission delay*. This delay is due to the limited bandwidth of the link on which the packet will be transmitted. Transmission delay calculates the time between the first and last bits of the packet being put on the wire. For example, a 500-byte (4 Kb) packet on a 1 Mbps link will incur a transmission delay of 4 ms because of constraints imposed by the bandwidth of the link.

Delay is an additive quantity. All types of delay incurred at all components add up. Thus, the longer the path, the more the number of routers that a packet will pass through, and the more delayed it is. Furthermore, if traffic at some other source increased so that the packet concerned sees a large queue, it will be delayed further.

1.3.2   Packet loss

There is no concept of loss in the circuit-switched networks. If a connection is established, then until the connection is terminated by the involved parties, all information communicated over the circuit will follow the established circuits and reach the other end. There will be no information loss.

In case of packet-switching, there is a possibility of packet loss. As discussed earlier, this happens in the extreme case where the buffer on a router's output interface is full and a new packet arrives. There is no space for the packet in the queue and hence it has to be dropped. Second, there is no notion of a circuit, so there is no notification to the involved parties that their packet was dropped. In fact, for reliable transmission of information in packet-switched networks such as the Internet, special protocols such as TCP have to be designed that identify a packet as being lost (somewhere along the path) and retransmit the packet so that the receiver obtains its content. While increasing packet delays serve as an indication that the queues in some routers are building up, the Internet protocols such as TCP react more drastically to a drop-in packet so as to reduce the load, thereby ameliorating congestion.

While a router's output interface queue becomes overloaded due to a surge in the traffic from some (potentially other) source , the impact of that surge is seen by our packet under consideration. This type of cross-interaction is possible due to packet switching.

1.3.3   Jitter

Jitter represents the variance in delay seen over a bunch of packets belonging to the same end-to-end connection. Simply put, over the life of a connection, several packets will be exchanged between the source and the destination. It is highly unlikely that each of these packets will face exactly the same queueing delay at all the routers along the path. In fact, given the Internet routing model, it is not guaranteed that all the packets will follow the same path and encounter the same routers.

This variability in the latencies of different packets of a connection is referred to as jitter. There is no jitter in a circuit-switched network. This is because once the end-to-end circuit is set up, there is no one contending with the corresponding connection to grab a share of that circuit.

From the perspective of VoIP, each packet carries some data corresponding to what was spoken. With a large jitter, the words that a packet contains would seem either too cluttered or too spread apart if the packets are played out as and when they arrive. To smooth out this effect, a *jitter buffer* is used to hold the packets for a while and release them at a smooth rate to the application to play.

1.4   QUALITY OF SERVICE (QoS) GUARANTEES

We have seen that in its native form, the Internet suffers from several problems that can have a significant impact on the performance of real-time applications such as VoIP. In order to counter these scenarios, the Internet Engineering Task Force (IETF), proposed mechanisms where the flows with such real-time requirements would be segregated from the other flows, even while they use the same router equipment. In essence, an application could request a certain amount of network resources along its entire path and its packets can receive preferential treatment from the network. Thus, the packets would be admitted in a special queue to conceal them from the effects of other traffic, and be scheduled for transmission with a higher priority. These requirements have been formalized by two standard mechanisms: Integrated Services and Differentiated Services.

In order to realize both these QoS models, the current Internet architecture needs to be altered, the service model changed, and the router functionalities modified to support the new services. We look first at the architectural changes that are required for both QoS services.

The QoS architectures could be classified broadly into two categories: *stateful* and *stateless*. The stateful architectures require per-flow states at all routers; the stateless architectures do not have such a requirement. In practice, the stateless architectures are actually core-stateless where the edge routers of a domain maintain per-flow state and the core routers do not. The major benefit of the stateless architectures comes from eliminating the costly packet classification operation at the core routers. The Integrated Services architecture is stateful whereas Differentiated Services is stateless in this terminology. In both architectures, various router functionalities are altered, and in turn they provide different levels of guarantee as well as having a different level of impact on scalability.

1.4.1   Integrated services

The Integrated Services (Intserv) framework intends to provide strong QoS guarantees to flows. Intserv requires that all routers have a per-flow state. Each router has a separate queue in the output buffer for each flow. A packet is added to the tail of its flow's queue in the output buffer. This adds another problem for the queue management component: If it has to drop a packet, then it has to take an additional decision about which queue's packet should be dropped. The scheduler is no longer FIFO because it has to select a queue from which to send the next packet. Since the number of queues is the same as the number of flows, the time complexity of the scheduler depends on the number of flows. The choice of a particular scheduler depends on the type of service provided under the Intserv purview. There are two key services defined under the Intserv framework: *Guaranteed and Controlled-load Services.*

- *Guaranteed Services:* Guaranteed service semantics intend to provide per-flow bandwidth and delay guarantees [1]. The routers have to ensure that its packets are never dropped as long as they are compliant with its traffic specification. Additionally, the scheduler employed has to schedule the packets of the flow based on its deadline and rate requirements. However, the complexity of these schedulers is significant and limits the scalability of the framework.

- *Controlled-load Services:* The controlled-load service intends to isolate a flow from the impact of other flows. The key specification of the controlled-load semantics is to provide an uncongested network view to a flow. The controlled-load service intends to provide a service similar to best-effort service when the routers are unloaded [2]. This type of service is suitable for adaptive real-time applications. VoIP is well-suited for this type of service.

In summary, the Intserv model has strong per-flow service semantics. However, it requires maintenance of per-flow states which renders it unscalable in the number of flows.

1.4.2   Differentiated services

As an architecture that does not mandate the per-flow state at all routers, Differentiated Services (Diffserv) is more scalable than Intserv. Diffserv classifies the routers as edge and core routers. Under Diffserv, only the edge routers maintain a per-flow state. On receiving a packet, an edge router classifies it to find the class it belongs to, and marks the type of service in the packet's header using a Differentiated Services Code Point (DSCP) that represents its class. The core routers only maintain a small number of queues corresponding to the number of classes and implement different *per-hop-behaviors* to service different DSCPs. They do not distinguish between individual flows and serve the packets having the same

DSCP in an identical fashion irrespective of the flow to which they belong. Since the core routers only have a fixed number of service classes (defined by the number of DSCPs), their scalability becomes independent of the number of flows.

The Diffserv framework has two types of defined service: *Assured Service and Premium Service.*

- *Assured Service:* Assured service aims at providing a lightly loaded network view by giving better-than-best effort drop rates to flows [3]. This is attained by implementing preferential dropping where a customer's *out-of-profile* traffic is dropped before his *in-profile* traffic. At the ingress router the user's packets are marked as in-profile or out-of-profile using a meter (or the user could indicate his preference). If a router becomes congested, it will drop the out-of-profile packets first. Thus, the user is assured of a fixed bandwidth (given by its in-profile rate).

- *Premium Service:* Premium service provides a virtual wire of a desired bandwidth from an ingress point to an egress point [4]. It is implemented using priority queuing to forward the premium packets at the earliest. Note that premium service can provide a bandwidth guarantee to the entire aggregate but since it does not distinguish between packets of individual flows, the delay bounds of individual flows cannot be distinguished, i.e. all flows in an aggregate have the same delay bound irrespective of their requirements.

Thus, the Diffserv framework alleviates the scalability problem of Intserv but can only provide weaker service semantics.

1.4.3   Other modifications

The QoS architectures require changes in the functionalities depending on the type of service model. They also require some additional operations which help in providing these services efficiently. We discuss these operations briefly.

1.4.3.1   Route pinning   One of the most fundamental changes required to support the QoS services is the ability to pin a flow to a fixed route. The conventional IP routing allows the routes to change at any time and different packets from the same session can take different paths. These changes could occur based on the underlying traffic changes, for load-balancing or due to topology alterations. To provide bandwidth or delay guarantees to a flow, the network has to be sure that the flow's path has sufficient resources to meet its requirements. If a flow's path changes during its lifetime, the new path might not have the desired resources.

Route-pinning techniques make sure that a flow follows its assigned path during its entire lifetime. The most prominent of these strategies are IP source routing, Multi-Protocol Label Switching (MPLS) [5], and Virtual Circuit Switching in ATM networks. In recent years, MPLS has become increasingly popular as a route-pinning and efficient packet-forwarding technology.

1.4.3.2   Packet classification   The forwarding mechanism in the Internet is based on the longest prefix match algorithm which takes the destination IP address of a packet as an input. The second change common to both QoS architectures is an ability to classify packets based on fields in their header other than the IP destination address. This is a must because the routers need to identify *which* packets belong to a flow with QoS guarantees and *how* these packets need to be treated. This requires a special packet classification functionality which essentially supersedes the routing lookup operation. Packet classification could be done based on multiple fields in the packet header, e.g. the source and destination addresses, protocol number and type of service field.

1.4.4 Admission control

Admission control refers to the process of limiting the number of QoS flows in the system so that their respective QoS guarantees are not violated. Specifically, on receiving a new request for some QoS guarantee, it is the responsibility of the admission control component to test whether there are sufficient resources in terms of bandwidth and buffer at the routers to support the new flow without violating the guarantees of the existing flows.

Of all the services listed above, only assured service could remain meaningful in the absence of admission control. All other services require some sort of admission control. However, the specific type of admission control they require varies, based on their respective characteristics. The admission control methods in the literature could be broadly classified as *deterministic* or *statistical* in nature. The deterministic admission control methods take as input the request parameters such as delay and bandwidth and determine whether or not the request could be supported after taking into account the existing reservations and using the knowledge of the scheduler characteristics and buffer availability. Statistical admission control methods on the other hand try to estimate whether or not the request could receive its desired service with some probability.

1.4.5 Status

The two QoS architectures for the Internet have been explicitly defined. However, they still await deployment in the Internet. This can be attributed to a lack of business model for these services along with the fact that any flow's path will traverse through multiple Internet Service Providers (ISPs) from source to destination. For a QoS guarantee to be given, all these traversed domains have to cooperate at the fine-granularity of per-flow or per-packet. This limitation is another reason for the lack of deployment of these services. Nonetheless, knowledge of these architectures is important as the techniques developed in this context are relevant in VoIP applications.

1.5 SUMMARY

Switching is a core concept that enables a telephony network to scale to a large user population. The traditional PSTN network uses a circuit-switching model where an end-to-end path is established prior to the communication starting. The circuit is dedicated to the call for its entire lifetime. The Internet uses a packet-switching model where the information to be communicated is packed into destination-labelled packets. Individual packets belonging to a single end-to-end session are switched independently. In both cases, routing is used to determine the end-to-end path – in PSTN the complete path is set up *a priori* whereas in the Internet, the next hop for a packet is determined when the packet arrives at a router. The addressing schemes in both types of network enable aggregation of addresses into more compact representations and thereby, reduction in the routing information.

The packet switching used in the Internet provides efficient resource utilization and scalability, but it results in additional performance problems in terms of packets facing excessive delays that could vary significantly from packet to packet and in the worst case the packet could be lost. In order to provide a certain quality of service guarantee to tackle these problems, two major standards have been defined that allow flows preferential treatment at the routers along their path. In order to ensure that the network has sufficient resources to support an additional request, admission control is used to determine whether there are sufficient resources remaining to address a request's demands.

REFERENCES

1. Shenker, S., Partridge, C. and Guerin, R. Specification of guaranteed quality of service. *IETF Request for Comments RFC 2211* (1997).

2. Wroclawski, J. Specification of the controlled-load network element service. *IETF Request for Comments RFC 2211* (1997).

3. Heinanen, J., Baker, F., Weiss, W. and Wroclawski, J. Assured forwarding PHB group. *IETF Request for Comments RFC 2597* (1999).

4. Jacobson, V., Nichols, K. and Poduri, K. An expedited forwarding PHB. *IETF Request for Comments RFC 2598* (1999).

5. Rosen, E., Viswanathan A. and Callon, R. Multiprotocol label switching architecture. *IETF Request for Comments RFC 2597* (2001).

CHAPTER 2

BASICS OF VoIP

This chapter gives an overview of the components, their interactions and their need in a VoIP deployment. This serves as a primer for a bird's eye view of a VoIP system – the desired functionality, the constraints, the components and an overview of how these components are tied together to provide end-to-end voice calls.

This chapter gives the reader the big picture of the VoIP arena. The goal is to have a frame of reference to understand where various components fit in when deploying a VoIP network. After understanding this chapter, the reader will be able to place the content of the more involved chapters in the correct context.

2.1 PACKETIZATION OF VOICE

Before venturing into the generic description of how VoIP works, we need to address a more fundamental question: how is the voice carried across the Internet? Since VoIP carries voice across the Internet, the mode of transport for the voice has to follow that of the underlying network. We saw in Chapter 1 that the Internet breaks the information to be transported across the network to a destination into small packets. The packets are then sent to the destination independently and the destination reassembles the desired information from a collection of packets. Hence, in order for VoIP to work, there is a need to packetize a speaker's voice. The voice content of a call has to be encoded into one or more packets, the packets should travel across the Internet like any other packets, and the receiver of those packets should be able to decode the voice content from the packets.

VoIP: Wireless, P2P and New Enterprise Voice over IP Samrat Ganguly and Sudeept Bhatnagar
© 2008 John Wiley & Sons, Ltd

Voice was transmitted in the form of analog signals in the old days of PSTN. The signals have to be digitized so that they become amenable to transportation in packets. This is done by sampling the analog waveform at 8000 times per second (Hertz). Then these samples are digitized, encoded using codecs, and multiple such samples are encapsulated into a single packet before being sent towards the destination. More details on the codecs and their performance will be discussed in Chapter 3. For the following discussion, the reader can assume that there is a way of converting voice to packets so that they can be transmitted and routed over the Internet.

2.2   NETWORKING TECHNOLOGY

The Internet contains diverse portions that are built upon different technologies. Like any other data traffic, VoIP also has to operate in these scenarios.

We described in Chapter 1 the basic form of network technology that the Internet uses. The basic form of Internet communication is comprised of packets traversing over routers. The performance issues that this architecture raises are known. However, a wide range of technologies with unique capabilities and limitations are emerging. They have their own impact on VoIP.

Conceptually we can think of the end-to-end path of a call as being comprised of several links. Some of these links are the traditional Internet links that behave in the manner we discussed earlier. The other links can belong to these emerging technologies. These could be the wireless 802.11 links that a VoIP phone can use to access the network or a link behind a NAT/Firewall in an enterprise that limits the type of communication possible. Similarly some portion of the network could be an overlay network resulting in more complicated address lookups than the Internet itself. Subsequent chapters deal with the impact of these technologies in detail.

2.3   ARCHITECTURE OVERVIEW

From an architectural standpoint, the minimum requirement to enable a VoIP call is to have two listening parties, each having a calling device equipped with a VoIP codec and connected over an IP network. However, as VoIP becomes a mainstream service with user demanding services that match and supersede the PSTN-level services, new functional components are being introduced into the VoIP architecture. Consequently, the current VoIP architecture is evolving rapidly by adding new services over VoIP and in addressing various issues specific to the deployment of VoIP over carrier networks, Enterprise LAN, etc.

Unfortunately, there exists no standardized VoIP architecture that can cover all the possible deployment scenarios and functionalities. Currently, different VoIP vendors and service providers have created their own unique architectures in order to differentiate themselves in terms of their functionalities. Yet, it is possible to refer to a generic VoIP architecture as shown in Figure 2.1 for discussing the functional requirements and associated functional components of a next-generation VoIP architecture.

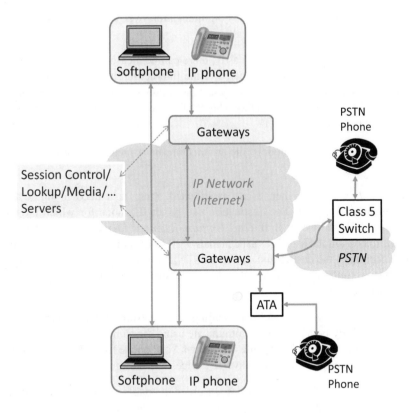

Figure 2.1   VoIP architecture.

2.3.1   Architectural requirements

The basic architectural requirements are derived from the deployment scenarios that enable a flexible communication model. Since the VoIP architecture is meant to enable voice calls over a packet-switched IP network such as the Internet, there are certain types of communication model that it must support. These can be listed as:

- *Internet-to-Internet:* This type of call includes those that originate on a phone connected to the Internet terminate at a phone connected to the Internet and the entire route remains inside the Internet.

- *Internet-to-PSTN:* These calls have the caller having a phone connected to the Internet whereas the callee is connected to the PSTN. Here the call traverses through both the PSTN segment and the Internet.

- *PSTN-to-Internet:* In this setting, the caller is connected to the PSTN whereas the callee has a phone connected to the Internet. Here, also, the call traverses both the PSTN segment and the Internet.

- *PSTN-to-PSTN via the Internet:* There is a case where the call originates and terminates on devices connected to the PSTN but the call's routing is done over the Internet. This can be done as communication over the Internet is cheaper and typically used for international calls.

- *Internet-to-Internet-via-PSTN:* Lastly, there can be a case where the call originates and terminates on devices connected to the Internet but a part of the call's route is over the PSTN. This can be the case when the circuit-switched link through the PSTN reduces the communication delay whereas the end-to-end Internet path may have a higher expected delay.

In order to support these models, the architecture must meet the following functional requirements:

- *Address Discovery:* When a call is initiated, there is a need to figure out the destination's location. The destination can be an IP phone for which the address may be an IP address or an Internet *Uniform Resource Identifier (URI)*. The address can also be a unique userID as used in many P2P VoIP applications. For supporting the PSTN phones, the destination can be a PSTN phone number. The address discovery service is important in any VoIP architecture for forwarding the call request to the appropriate entity.

- *Device Interoperability:* A VoIP calling device from different vendors should interoperate by being able to communicate using the same protocol. A VoIP phone from vendor A should be capable of calling a VoIP phone from Vendor B. Following the standards ensures that such diverse devices remain interoperable.

- *Interoperability with PSTN phones:* In order to enable calling to and from PSTN phones, the architecture must provide functionalities that provide protocol level translation and VoIP data level transcoding. With these functionalities, the call generated from the IP network can be forwarded to and from PSTN network class 5 switches.

- *Session-level Control:* In different deployment scenarios, various session-level control functionalities become important. Such functionalities include session-level authorization, authentication, user billing, etc.

- *Media-level Functionalities:* Media-level functionalities refer to services provided to the actual voice over data that is transported over media transport protocols such as RTP. Such functionalities include various media-level processing to enable call mixing for multiparty conferencing, transcoding to enable transport over heterogeneous network links, etc.

- *Interoperability among Components:* All the functional components of a VoIP architecture should be interoperable by using standard protocols (such as SIP/H.323). This will enable (a) multivendor equipments to inter-operate; and (b) multiple VoIP service providers to coordinate in carrying each others' VoIP traffic.

The requirements listed above are not exhaustive. They merely represent the high-level requirements that are required for VoIP. Fortunately, these requirements are well addressed by the recently proposed VoIP architectures at different levels. Each of these high-level requirements may lead to various deeper level functionality requirements that are beyond the focus of this section.

The next step in understanding the VoIP architecture is to define the minimum set of functional components and protocols that are required to meet the above requirements.

2.3.2 Functional components

Having defined an abstract architecture for VoIP, we now look at some of the functional components that this architecture needs.

2.3.2.1 VoIP calling device   These are the devices that an end-user requires to initiate or receive a call. These include:

- *IP Telephone:* An IP telephone is a device which can directly be connected to the Internet. It has some in-built software that allows it to communicate with other VoIP devices. Along with other features, this software can provide the functionality to set up a call and the protocols necessary to transport the voice packets. An IP phone can connect to a network using a standard RJ-45 Ethernet connector or it can be a wi-fi VoIP phone that can connect to the Internet using the IEEE 802.11 wireless networks.

- *Softphone:* A softphone is a telephone implemented entirely in software. The softphone runs on a computer or a PDA and a user can use it to dial any number such as in traditional telephony. VoIP does not distinguish between a softphone and a hardphone. In fact, this allows the user to have no additional telephone equipment and makes computer-to-computer or computer-to-PSTN calls possible.

- *Analog Telephone:* An analog telephone is one that is traditionally used to connect to the PSTN.

- *Analog Telephone Adapter (ATA):* If a user wishes to use a legacy analog telephone to connect to the Internet, an ATA can be used. A stand-alone ATA contains the logic to communicate with the service provider over the Internet and to translate the communication to and from the telephone.

Any VoIP-capable calling device must contain the VoIP codec whose purpose is to encode digitized samples of voice and make them amenable to transmission as packets. More details on codecs are available in subsequent chapters.

2.3.2.2 Gateway   Perhaps the most important component to establish the viability of a VoIP system is a gateway. The task of a gateway is to sit at the border of two different types of network and help them communicate. In the case of VoIP, these two networks can be the Internet and the PSTN. Typically, a gateway consists of two main components: (a) a gateway controller; (b) a media gateway.

The gateway controller is responsible for the following roles: (a) to translate the information into a format that each network can understand; (b) to enable inter-operation of the signaling. For example, when supporting PSTN, the gateway controller can translate the SIP signaling information for a VoIP call into the equivalent SS7 signaling information over PSTN. This translation allows a call request at the IP network to be forwarded to the PSTN and *vice versa*. Enabling inter-operation of the signaling is important in the divided world of VoIP where both SIP and H.323 coexist for session-level signaling. For example, the majority of the carrier networks still use H.323 while a major portion of the IP phones used inside enterprises use SIP. A gateway controller can act as an intermediary for translation between SIP and H.323.

A media gateway is a component associated with the gateway controller and performs similar tasks but on the media level. A media gateway performs media transcoding. For example, before forwarding the VoIP data from IP networks to the PSTN, the media gateway provides transcoding from the packet-based VoIP in the IP network to the corresponding frames in the TDM network of the PSTN. The transcoding function can be used when forwarding VoIP from one IP network to another, particularly when there is a difference in the supported bit rate or speech-coding type.

2.3.2.3 Media server   In certain deployment architectures, the media gateway functionalities can be augmented by another component called media server. The media server has the role of processing the VoIP RTP stream for providing Dual-Tone Multi-Frequency (DTMF) tone decoding, mixing multiple media streams into a conference, playing announcement, processing VoiceXML scripts, speech recognition, text to speech conversion, audio recording, etc. It is, however, unnecessary to have a separate media server; its role can be integrated into the media gateway.

2.3.2.4 Session control server   Session control server has the general role of providing session-level functionalities such as authentication, authorization and admission of VoIP calls. The same role can support call routing and forwarding to another network or service provider and can maintain states about ongoing calls. There are other auxiliary roles in providing functionalities such as caller ID, call waiting or interaction with application servers. Session control server is also an optional component in a VoIP architecture and can be integrated as part of the gateway controller. Session control servers are also referred to as SIP servers or call agents depending on the context.

2.3.3 Protocols

The protocols define the communication messages, their meaning and the corresponding logic. Protocols are required for several purposes in VoIP.

1. *Address Discovery* – These protocols provide the support for discovering the destination's location when a call is initiated.

2. *Call Signaling* – These protocols perform the task of setting up and tearing down a call between the caller and the callee.

3. *Gateway Control* – These provide the signaling necessary to control the functionality of a gateway.

4. *Media Transport* – The previous two protocol groups represent control functionality. The actual voice is transmitted over the media transport protocols.

We shall discuss a few important protocols in Chapter 5.

2.4 PROCESS OF MAKING A VoIP CALL

Using the aforementioned functional components and protocols, we can trace a call through a simple scenario in our architecture. When a user needs to call a destination, it contacts a gateway to help look up the number. The location of the gateway itself may vary, depending on the specific VoIP architecture and protocols involved. The lookup service and protocols are used to determine the call routing for the desired call. This may include finding the shortest or the least-cost path through both the PSTN and the Internet. The source gateway may then contact the destination gateway to negotiate the parameters of the call using some signaling protocols. Next, the media path for the two endpoints is set up that may go through a media gateway to provide appropriate translation.

While this procedure is relevant to the generic architecture we propose, at the core of any architecture, this is the required functionality. These architectures may vary in the specifics of how and where each functional component is implemented and how the actual interaction takes place. Furthermore, they have to deal with several practical issues such as authentication, security and handling diverse devices such as NATs and firewalls.

2.5 DEPLOYMENT ISSUES

There are multiple types of deployment scenario for VoIP service. Presently, the VoIP deployment scenarios can be categories into the following types:

- P2P VoIP over Internet;

- Managed VoIP over Internet;

- VoIP over managed carrier IP networks;

- VoIP over Enterprise networks.

Each type of deployment scenario has its own unique set of issues. In the remaining part of the book, we discuss many such issues in detail by focussing on each of the above deployment scenarios. In this section, a basic overview of the issues is provided.

P2P VoIP deployment started with the era of Skype as the popular application for VoIP communication. The main issues that Skype addressed are: (a) how to locate a user from his/her user-Id; (b) how to support NAT traversal. Locating a user in a P2P network can be done by using Distributed Hash Table (DHT) or variants of the same technology. The main problem tackled by DHT is to provide a scalable and distributed approach to track users without maintaining a central server hosting a mapping from the user-ID to his IP address. More details about DHT are discussed in Chapter 7. The subsequent chapters will detail the impact of overlay networks on VoIP.

The need to support NAT traversal has become a general problem in most VoIP deployments. Skype provided their proprietary solution for supporting NAT. However, there exist open solutions such as STUN for NAT traversal. Details of NAT traversal are discussed in Chapter 15.

In many ways, P2P VoIP deployment addressed the main issues and went on to establish the viability of VoIP by having a satisfied user-base numbering millions. At the same time, the second type of scenario became popular with the VoIP services from Vonage and later by cable providers such as Comcast. The difference in this deployment scenario was that it was a managed service and did not involve the peer/user to aid in any way, including location tracking. Further, the Vonage type of deployment addressed new issues such as: (a) supporting PSTN phone using ATA; (b) calling to/from PSTN numbers.

The third deployment type can be classified into two categories: (a) VoIP between PSTN networks; (b) direct VoIP to users. Case (a) existed for a long time where two carrier networks would transport PSTN calls using VoIP over IP networks. In order to support this, the carrier network would aggregate PSTN calls, transcode them into VOIP calls and deliver the VoIP over an IP network. This was an inexpensive solution for a service provider to support international calls. Case (b) is emerging slowly, where the carrier networks have realized that they can directly support VoIP to end users. The majority of the current evolution of the VoIP architecture is motivated by the interest of the carrier networks. The main issues being addressed in this scenario include methods to provide call routing and forwarding from one network to another and supporting new applications over VoIP.

The fourth deployment scenario is already widely popular among many enterprises. Typically, any enterprise setting allows two types of deployment case. In the first case, the VoIP resides inside the company network and uses an IP-PBX for VoIP call switching. All VoIP calls exit through PSTN lines. In the second case, the VoIP does not terminate at the boundary of the enterprise network or LAN, but rather, continues over the wide-area IP network to its destination. Supporting VoIP across network boundaries involves issues such as security, reliability, privacy, etc. We look at these issues in Chapter 17.

2.5.1 VoIP quality and performance issues

Running on top of heterogeneous IP networks while dealing with the distinct characteristics of each of them, and going through transcoding at the gateway of network boundaries causes significant performance issues that any VoIP deployment has to face. VoIP is susceptible to the underlying network conditions (delay, jitter and packet loss), which can degrade the voice application to the point of being unacceptable to the average user. We highlighted some of these issues in the context of generic Internet performance issues. We elaborate their impact on VoIP here.

2.5.2 Delay

Delay is the time taken by a packet to travel from one point to another (one-way) in a network. It is easy to measure the round-trip delay of the Internet. The performance of codecs differs due to their ability to tolerate delay, but a good rule of thumb is to limit the one-way delay to about 150 ms. VoIP packet delay is comprised of the following components:

- *Propagation Delay:* This delay is proportional to the speed of light and depends on the physical distance between the two communicators. A call traversing continents would face a significant amount of propagation delay.

- *Transport Delay:* Transport delay occurs because of network devices such as routers, firewalls, traffic shapers, etc. This delay includes the queueing delay and the packet-processing delay at each network point, and can vary with the traffic. Typically, the sum of round-trip propagation and transport delay for the Internet is approximately 90 ms.

- *Packetization Delay:* This is a function of the codec speeds. Low-speed codecs, such as the G.723, take approximately 67.5 ms to convert analog signals into digital packets. The extra time is required because these codecs have to compress the packets to reduce their size. High-speed codecs such as the G.711 can do the packetization in approximately 1 ms. We shall look at further details of packetization delay in Chapter 3.

- *Jitter Buffer Delay:* A jitter buffer helps to minimize the variations in the arrival times of the voice datagrams. However, sometimes in the event of excessive delay, packets have to be discarded.

2.5.3   Jitter

Jitter is the variation in delay over time. Jitter occurs due to different packets facing different amounts of delay in the Internet. In an overloaded network, buffering at intermediate routers results in an increase in jitter. If the jitter is high or delay of transmissions varies too widely during a VoIP call, the call quality goes through significant degradation. The amount of jitter tolerable on the network is affected by the depth of the jitter buffer on the network equipments in the voice path. The greater the size of the available jitter buffer, the more the network can reduce the effects of jitter. However, to maintain the high interactive level in VoIP, the jitter buffer cannot be very large.

2.5.4   Packet loss

Packet loss is a result of losing packets along the data path, thus severely degrading the voice application. Packet loss happens when the router queue/buffer limit is exceeded during congestion. In a wireless network, packet loss can happen due to bad channel conditions where both retransmission techniques and packet error recovery techniques fail successfully to deliver the packet to the receiver. Packet loss can also happen at the jitter buffer if the jitter is too high for the packet or the end-to-end delay for the VoIP packet is not met.

2.5.5   Echo and talk overlap

This effect, which consists of hearing a delayed repetition of the voice signal, can be encountered if there are more than 50 ms of round-trip delay. In order to avoid this undesired effect, codecs must implement some echo-erasers. The echo suppression techniques typically preserve the signals for a while and then subtract it from the subsequent signals containing the echo.

2.5.6   Approaches to maintaining VoIP quality

Maintaining VoIP quality can be addressed by two independent approaches: (a) design networks better to handle VoIP by enabling QoS provisioning strategies; (b) design better voice codecs and/or VoIP client to provide robustness to variation in network conditions.

2.5.6.1   Network-level QoS   There exist various QoS control mechanisms for the Internet as described in the previous chapter. Unfortunately, in the current setting, these mechanisms are not enabled in most of the IP routers that make up the Internet. Therefore, there are no end-to-end QoS guarantees provided to date. However, the current situation is slowly improving. Next generation wireless access networks such as WiMAX are providing QoS control mechanisms to support VoIP. Architectures such as IP Multimedia Subsystem

(IMS) are being proposed to support QoS. Even wireline access networks are also being geared towards providing some level of QoS to the end-user.

At the same time, there are various higher layer transport mechanisms being proposed to circumvent the lack of QoS in the Internet. To that end, overlay network solutions that can reroute calls with lower loss or delays can be effective in providing good quality end-to-end VoIP paths. More details about VoIP in overlay networks are provided in Chapter 8.

2.5.6.2 VoIP codecs The quality of VoIP can be significantly improved by using proper codec. As opposed to traditional narrowband codecs, current wideband codecs provide natural and crisp sound quality. New mechanisms are also being used by current VoIP codecs that can deal with packet loss and jitter. Multirate codecs operating at multiple bit rates are available today that adapt to network congestion or variation in the available bandwidth. Success of Skype using the Global IP Sounds (GIPS) codec proved that millions of users can be satisfied with the performance of VoIP over the Internet. However, it does not mean an end to the development of new codecs. Different networks with heterogeneous and unpredictable behavior are still challenging the VoIP codec designer to offer *better than PSTN* voice quality over IP networks.

2.6 VoIP APPLICATIONS AND SERVICES

The primary purpose of VoIP is to support voice communication between two parties, but the potential applications of VoIP are not restricted to that. It is the flexibility to add various services to VoIP (because of being an IP-based application) which is driving the widespread adoption of VoIP by different service providers. These services include those provided by PSTN and emerging ones that are not available with PSTN.

2.6.1 Fax

Fax is the most common application that VoIP supports. Supporting fax has been a challenge in VoIP networks as fax is very sensitive to packet loss. Also, low-bit rate codecs optimized for VoIP are not suitable for sending fax. However, recently, new protocols are being proposed to send fax directly over IP networks or via more reliable transport.

2.6.2 Emergency numbers

Supporting emergency numbers is difficult as the location of the user cannot be easily tracked in a VoIP network. It is therefore difficult both in terms of forwarding the call to the right authority and tracking the original location of the call. However, recent advancements in the area of GPS-based location services along with fixed IP addresses to location mapping are making it easier to support emergency numbers.

2.6.3 Roaming

Due to the IP-level communication, the exact location of the user is not important in placing a call. This inherent feature provides a user with the flexibility to avail the same service while away from home. For example, with Vonage, the user can easily use the service while visiting any location across countries.

2.6.4   Voice over IM

Voice over IP allows an Internet Messenger (IM) user to make VoIP calls from his IM inter-face. This application shows the ease of bundling the VoIP service with other applications.

2.6.5   Push-to-talk

Push-to-talk is easily supported by VoIP as, by being non-interactive, it does not require the same level of QoS guarantee. Push-to-talk also does not require big changes in the VoIP architecture. Push-to-talk is equivalent to sending a streaming audio media over the same VoIP connect.

2.6.6   Conferencing

A traditional two-party voice call can be turned into a multiparty conference very easily. At a high level, conferencing can be enabled by replicating the voice stream emanating from the user to all others participating in the conference. The process of replication can be optimized by using various techniques based on the available bandwidth of different users. In general, a user with a higher available outgoing bandwidth can perform more replications than one with a low outgoing bandwidth.

2.6.7   Integration with other applications

Again by being an IP-based solution, VoIP can be easily integrated with other applications. For example, a VoIP can be integrated with the e-mail application to convert a voice message to e-mail and send it to the receiving party or vice versa. Such integration enables new ways of using VoIP beyond just voice communication.

2.7   SUMMARY

This chapter illustrates the fundamental requirements that are being met in VoIP delivery for widespread deployment and mass acceptance of VoIP. These requirements address various issues, among which the most important are:

- interoperability among multiple networks (IP networks and PSTN) and service providers;
- maintaining the QoS across heterogeneous IP networks with unpredictable perfor-mance;
- auxiliary PSTN equivalent features that the user expects.

Current evolution in the VoIP architectures, VoIP codec and software design and new techniques such as NAT traversal are addressing the above requirements.

CHAPTER 3

VoIP CODECS

VoIP services are steadily being deployed over diverse types of packet-switched IP network that include Internet, Enterprise LAN and wireless networks. The core component of any VoIP software is the codec, that determines how the voice/speech samples are encoded into packets and transmitted over the network. Today, there exists a wide range of codecs to match various performance requirements and network conditions. It is important to understand the design of various codecs as that guides the codec selection for a given VoIP deployment scenario. This chapter provides a basic understanding of the popular codecs in terms of their design and features.

3.1 CODEC DESIGN OVERVIEW

Voice codec is the most critical component of a VoIP system. The primary role of a voice codec is to convert the input speech signal into digital form, transmit the signal to the receiver and reconstruct the original speech signal for the listener. The end-user perceived voice quality strongly depends on the performance of the codec in terms of the accuracy in reproducing the speech at the receiver end.

The performance of a codec is closely tied to the techniques followed by the codec in the process of voice transmission. The process of transmission of the human voice follows three steps:

VoIP: Wireless, P2P and New Enterprise Voice over IP   Samrat Ganguly and Sudeept Bhatnagar
© 2008 John Wiley & Sons, Ltd

- voice sampling;

- quantization;

- voice coding.

The codec samples the waveform at regular intervals and generates a value for each sample. These samples are typically taken 8000 times/s (8 kHz sampling rate) or 16 000 times/s (16 kHz sampling rate). The sample values are then quantized in order to map values into discrete-finite values that can be represented using bits. In the coding step, the samples are accumulated for a fixed period of time and encoded into a set of bits. Encoding is a process of information compression where the number of bits to represent the samples is minimized. This set of bits forms the voice data frame that is subsequently transmitted over the network. In the overall process, it is the last step of coding that determines the efficiency of the codec in terms of bandwidth requirement.

The primary goal of speech coding is to minimize the bit rate in the digital representation of a signal without an objectionable loss of signal quality in the process. Along with meeting the above goal, the speech-coding process must also have low complexity and low packetization delay. Complexity refers to the time and space complexity of the coding algorithm. The packetization delay (also referred to as algorithm delay) refers to the maximum waiting time of a given voice sample until the voice frame or packet is created that contains the sample.

The underlying approach used in any speech-coding technique for minimizing the bit rate is to exploit signal redundancy as well as the knowledge that certain types of coding distortion are imperceptible because they are masked by the signal. Models of signal redundancy and distortion masking are becoming increasingly sophisticated, leading to continued improvements in the quality of low bit rate codecs.

In the past, significant research and development with the above design approach has resulted in high-quality voice codecs that require very low bandwidths for transmission. These codecs were designed for the PSTN network and were standardized by the International Telecommunication Union (ITU-T) as G.711 and G.72x series of codecs. PSTN lines provided low bandwidths that motivated the design goal of low bitrate voice communication. As a result, the codecs developed for PSTN fall under the class of narrowband codecs. These narrowband codecs are still widely popular, not only in providing voice over PSTN but also in providing voice over the wireless cellular network.

3.1.1 VoIP codec design goals

In recent times, with the deployment of VoIP over the Internet, Enterprise LAN and Broadband Wireless Networks, the bandwidth availability has become less of a problem. Rather, what became questionable to the service provider is the applicability of the narrowband codecs for packet-switched IP networks. It was not well understood if the existing narrowband codecs originally designed for circuit-switched PSTN can handle the packet loss, large delay and jitter that exist in any IP network. Clearly, just minimizing bit rate in a codec does not guarantee its acceptability for supporting VoIP. Instead, a VoIP codec has to be designed such that the codec is robust to all the network-level problems and offers a smoother-quality degradation to underlying change in network conditions.

Over the past few years, considerable progress in the design of VoIP codecs has been made to address the above problems. Increased deployment and usage of VoIP over the

Internet and enterprise LAN has proved the initial point that the current VoIP codecs are acceptable to the users in terms of the quality of experience. The new trend of VoIP has also driven the design and use of the new class of wideband codecs and adaptive multirate codecs.

Wideband codec-based VoIP offers a more natural and crisp-sounding voice that is not experienced in the voice communication over PSTN. Adaptive multirate codecs are specifically designed for packet-switched networks such as the Internet to provide the ability to adapt to the underlying network congestion. Furthermore, different techniques were incorporated in the VoIP codecs such as jitter buffer adaptation and speech gap filling in order to adapt to the jitter and packet loss problems.

With all these advancements in VoIP codecs and better QoS management capabilities in the IP networks, the future VoIP is poised to provide better than PSTN experience to the users.

3.2 SPEECH CODING TECHNIQUES

The speech coding techniques were designed for application in circuit switched networks such as PSTN or wireless cellular network. However, the concepts developed for the speech coding are fundamental and apply beyond circuit switched networks. As a matter of fact, the new class of VoIP codecs also leverage the coding techniques initially designed for PSTN. Therefore, it is important to understand the classes of speech coding technique and their evolution in terms of design.

Depending on the speech coding and transmission techniques used, the existing codecs can be categorized into four broad classes:

- waveform codecs;

- source coding;

- hybrid coding;

- multirate codecs.

Typically, waveform codecs provide a very high-quality speech but at the cost of using higher bit rates. On the other hand, source codecs operate at a very low bit rate but produce a more synthetic speech reproduction. A combination of both the above techniques is employed in hybrid codecs. Hybrid codecs result in good quality speech while maintaining lower bit rates. Most existing popular codecs are hybrid codecs. While multirate codecs are based on the above speech coding techniques, they can operate on multiple bit rates. Multirate codecs are better equipped to adapt to the underlying network resources such as variation in terms of the available bandwidth. A detailed description of the speech coding techniques for the above classes of codec is presented next.

3.2.1 Waveform codecs

Waveform codecs operate on the basic principles of speech coding where the input speech signal is converted into digital signal and subsequently packetized. The goal of waveform codec is to produce a reconstructed signal at the receiver as close as possible to the original one while reducing the bandwidth requirement. The codecs based on the above techniques are generally differentiated in terms of the coding complexities. Low-complexity codecs perform well at higher bit rates. However, the quality degrades rapidly when the data rate is lowered to below a level that is typically around 16 kb/s. Some of the basic techniques that underly waveform codecs are as follows.

3.2.1.1 *Pulse code modulation (PCM)*

Pulse code modulation (PCM) [1] is the simplest form for waveform coding where sampling is done at 8000 samples/s. If a simple linear quantization is used requiring 12 bits per sample, the resultant bit rate is around 96 kb/s. This rate can be easily reduced by using a nonlinear or logarithmic quantizer with 8 bits per sample resulting in a bit rate of 64 kb/s. Such a logarithmic quantizer can also provide good quality over a wide range of input levels. Two such logarithmic quantizer-based PCM codecs were standardized in the 1960s. In America, μ-law coding is the standard, while in Europe the slightly different A-law compression is used. Because of their simplicity, excellent quality and low delay, both of these PCM codecs are still widely used today.

3.2.1.2 *Differential PCM (DPCM)*

With the aim of increasing the efficiency of PCM, subsequent research resulted in the development of Differential Pulse Code Modulation (DPCM) [2]. The underlying technique in DPCM is based on predicting the value of the next sample from the previous samples. Such a prediction is possible due to the observation that there exists a high level of correlation in speech samples. The presence of such correlation is due to the effects of the vocal tract and the vibrations of the vocal cords. Using prediction, the number of bits necessary to represent a set of voice samples is minimized as one can just quantize the difference between the original and the predicted signals. On the receiver side, the signal is rebuilt by adding the differential value to the previous one to obtain the predicted signal value.

The results from such codecs can be improved even further if the predictor and quantizer are made adaptive so that they change to match the characteristics of the speech being coded. In other words, the amplitude of the quantization intervals is dynamically changed, but the number of quantization levels is kept fixed. The result of such improvement led to the development of Adaptive Differential PCM (ADPCM) codecs. In the mid-1980s the CCITT standardized an ADPCM codec operating at 32 kbits/s, with resulting speech quality almost equal to the 64 kbits/s PCM codecs. Later ADPCM codecs operating at 16, 24 and 40 kbits/s were also standardized.

3.2.2 Source coding

The operation of source coders is based on employing a model-based representation of the speech signal. Therefore, instead of sending the actual waveform, the parameters of the model are transmitted. The receivers construct the actual waveform from these parameters. The model is built by representing the vocal tract as a time-varying filter. The filter is excited with either a white noise source representing unvoiced speech periods or a train of pulses separated by the pitch period for voiced speech. The transmitter sends the decoder the complete filter specification, a voiced/unvoiced flag, the necessary variance of the excitation signal, and the pitch period for voiced speech. The information is sent every 10–20 ms to capture the time-varying non-stationary nature of the speech.

The codecs used for source coding are also called *Vocoders*. An example of such a codec is Linear Predictive Coding (LPC) [3]. Vocoders can operate at extremely low bit rates of around 2.4 Kb/s. Vocoder-produced speech is comprehensible but does not sound natural. The performance cannot be improved by increasing bit rate due to the limitation of the technique.

3.2.3 Hybrid coding

The most popular and efficient codecs are designed based on leveraging the benefits of waveform and source coding. Hybrid codecs attempt to fill the gap between waveform and source codecs. The simple form of codec based on hybrid coding principles is the Analysis-by-Synthesis (AbS) codec. The AbS codec does not just apply a simple two-state voice/unvoiced model to find the necessary input to the filter. Instead, an AbS codec works by splitting the input speech to be coded into frames, typically about 20 ms long. For each frame, parameters are determined for a synthesis filter, and then the excitation to this filter is determined. The excitation signal is determined to minimize the error between the input speech and the reconstructed speech. In other words, the encoder analyzes the input speech by synthesizing different approximations to it and then choosing the one that minimizes the error in approximation. After synthesis is completed, the encoder transmits for each frame the information representing the synthesis filter parameters and the excitation to the decoder. At the decoder, the given excitation is passed through the synthesis filter to generate the reconstructed speech.

Given the basic design of the AbS hybrid codec, subsequent developments were focused towards efficient representation of the excitation signal in order to reduce the amount of bits required to transmit the information and the complexity in creating the information.

The developments resulted in two types of codec: (a) Multi-Pulse Excited (MPE) and (b) Regular Pulse Excited (RPE). The MPE codecs represented the excitation signal for every frame of speech by a fixed number of non-zero pulses with associated amplitudes. The transmitted information consists of the position and amplitude of each of the pulses. However, the complexity in determining the pulse position and amplitude is high. In the case of RPE, the non-zero pulses are at a fixed interval. Therefore, only the position of the first pulse and amplitude of each pulse needs to determined and transmitted. This results in less information being transmitted for RPE. Both MPE and RPE can provide good-quality speech at rates of around 10 Kb/s and higher. However, both MPE and RPE cannot operate well below 10 Kb/s due to the large amount of information that must be transmitted corresponding to each pulse in a given voice frame. An example of RPE codec is the popular GSM codec used in the cellular networks.

The next level of optimization was proposed in the development of Code Excited Linear Prediction (CELP). In contrast to MPE and RPE, the excitation signal is vector quantized and is represented as an entry from a large vector quantizer codebook with an associated gain. Typically, code book index is represented with 10 bits and gain value with 5 bits; so, compared to RPE requiring 47 bits, CELP requires 15 bits to represent a voice frame. However, CELP also incurs a high amount of complexity in determining the right codebook entry. In CELP, the synthesis method has to go through a large number of entries to determine the appropriate one that minimizes the error. To that end, various CELP-based codecs were developed that improved on the basic CELP technique and minimized the complexity and the bit rate. Some of these CELP-based codecs include the low delay CELP, Department of Defence (DoD CELP) [5] and Conjugate-Structure Algebraic-Code-Excited Linear Prediction (CS-ACELP) [4]. The codecs provide a wide range in bit rate selection starting from 4.8 Kb/s to 16 Kb/s.

3.2.4 Adaptive multirate

AMR [6] speech codec represents a new generation of coding algorithms that have been developed to work with inaccurate transport channels. The flexibility on bandwidth requirements and the tolerance in bit errors of AMR codecs are not only beneficial for wireless links, but are also desirable for VoIP applications.

3.3 NARROWBAND CODECS

Based on the different speech coding techniques, ITU-T led the standardization of the class of narrow band codecs under the G.711 and G.72x series codec names. These standard speech codecs use different speech coding techniques and offer a wide range of bit rates, coding complexity and quality. Given this broad spectrum of codecs, it is important to understand the differentiating features of each codec to guide the appropriate selection matching the deployment scenario. Characteristic features of some of the popular codecs are presented in the following.

3.3.1 PCM-based G.711

G.711 codecs based on PCM were among the first standardized codecs. Typically, G.711 codec operated at various bit rates: 64 Kb/s (standard), 56 Kb/s and 48 Kb/s (non-standard). Because of their simplicity, excellent quality and low delay, both of these codecs are still widely used today. G.711 is also the mandatory minimum standard for all ISDN terminal equipment. G.711 Vocoder operates at the following bit rates: 64 Kbps (standard), 56 Kbps and 48 Kbps (non-standard).

3.3.2 ADPCM-based G.721 codecs

In the mid-1980s, the G.721 operating at 32 kb/s using ADPCM was standardized. G.721 is able to reconstruct speech almost as well as the 64 kbits/s PCM codecs. Later, in recommendations G.726 and G.727, codecs operating at 40, 32, 24 and 16 kbits/s using the same underlying technology were also standardized.

3.3.3 RPE-based GSM codec

The 'Global System for Mobile communications' (GSM) is a digital mobile radio system which is extensively used throughout Europe, and also in many other parts of the world. The GSM full rate speech codec operates at 13 kbits/s and uses an RPE codec. The GSM codec provides good-quality speech, although not as good as the slightly higher rate G.728 codec. The main advantage of GSM codec over other low rate codecs is its relative simplicity. GSM codec can easily run in real time on a 66 MHz 486 PC for example, with small data memory requirement. In contrast, CELP-based codecs require a dedicated DSP to run in real time.

3.3.4 Low-delay CELP-based G.728 codec

At bit rates of around 16 kbits/s and lower, the quality of waveform codecs falls rapidly. Thus at these rates, hybrid codecs (especially CELP codecs and their derivatives) tend to be used. However, as noted above, the CELP-based codecs tend to have high delays. The delay of a speech codec is defined as the time from when a speech sample arrives at the input of its encoder to when the corresponding sample is produced at the output of its decoder, assuming the bit stream from the encoder is fed directly to the decoder. For a typical hybrid speech codec this delay will be of the order of 50 to 100 ms, and such a high delay can cause problems. In 1988, the CCITT released a set of requirements for a new 16 kbits/s standard, the chief requirements being that the codec should have speech quality comparable to the G.721 operating at 32 kbits/s ADPCM codec in both error-free conditions and over-noisy channels, and should have a delay of less than 5 ms and ideally less than 2 ms.

All the CCITT requirements were met by a backward adaptive CELP codec which was developed at AT&T Bell Labs, and was standardized in 1992 as G.728. G.728 came to be known as the low-delay CELP codec. The codec operated at 16 kb/s with a delay of less than 2 ms with speech quality equal to or better than G.721 and a good robustness to channel errors.

3.3.5 DoD CELP-based G.723.1 codec

In 1991 the American Department of Defense (DoD) standardized a 4.8 kbits/s CELP codec as Federal Standard 1016. Based on the DOD CELP, the G.723.1 codec was standardized by ITU-T to operate at two bit rates: 6.3 Kb/s and 5.2 Kb/s. The 6.3 Kb/s bit-rate version uses a 24 byte frame and uses an improved DoD CELP algorithm called Multi-Pulse Maximum Likelihood Quantization (MP-MLQ). The 5.2 Kb/s version uses 20-byte frames using an improved DoD CELP algorithm called Algebraic Code Excited Linear Prediction (ACELP). G.723.1 codec compresses voice audio in 30 ms frames and requires an extra 7.5 ms lookahead for frame construction. Therefore, the total delay in generating a voice sample is 37.5 ms.

G.723.1 is useful for VoIP applications over bandwidth constraint links such as wireless networks. G.723.1 falls under the class of narrowband codec with 8000 samples/s sampling frequency.

3.3.6 CS-ACELP-based G.729 codec

ITU-T recommendation G.729 annex A (referred to as G.729A) resulted in the G.729a codec that uses Conjugate-Structure Algebraic-Code-Excited Linear Prediction (CS-ACELP). G.729 using CS-ACELP has lower complexity and consequently lower algorithmic delay. G.729A operates on 10 ms frames with 5 ms lookahead delay resulting in a total algorithmic delay of 15 ms. G.729A operates at 8 Kb/s with a sampling rate of 8000 samples/s and thereby falls into the class of narrowband codecs. Therefore, the G.729A is useful for VoIP applications over bandwidth-constrained links. An extended version of G.729 referred to as G.729B also includes silence compression techniques by using voice activity detection (VAD) to reduce the transmitted bit rate during the silence periods.

3.3.7 iLBC

iLBC (internet Low Bit-rate Codec) is a royalty free narrow band codec developed by Global IP Sounds (GIPS). iLBC was specifically designed for VoIP application and is based on a block-based Linear Predictive Coding algorithm. The iLBC codec enables graceful speech quality degradation in the case of lost frames, which occurs in a connection with lost or delayed IP packets. The codec operates at 13.33 Kb/s with an encoding frame length of 30 ms and at 15.20 kbps with an encoding length of 20 ms. iLBC has been observed to provide better quality than G.729A codec in terms of robustness to packet loss.

The complexity is similar to that of G.729A. iLBC is a widely popular codec and is used by softwares such as Googletalk, Skype and Gizmo Project.

3.3.8 Comparison of narrowband codecs

For a comparison of the basic characteristics of the above codecs for quick reference, see Table 3.1.

Table 3.1 VoIP codec comparison.

Codec	Technology	Bit rate (kb/s)	MOS	Complexity
G.711	PCM	64	4.1	Low
G.721	ADPCM	32	4.0	
G.723.1	ACELP/MP-MLQ	5.3/6.3	3.65	High
G.726	ADPCM	16/24/32/40	4.0	
G.728	LD-CELP	16	3.61	
G.729	CS-ACELP	8	3.92	Medium
iLBC	LPC	13.33	N/A	Medium

3.4 WIDEBAND AND MULTIRATE CODECS

The development and application of wideband codecs is mostly motivated by the deployment of VoIP over current broadband access networks where available bandwidth is high. Wideband codecs have a higher sampling rate (16 kHz) and provide better sound quality. Some of the wideband codecs used in current VoIP softwares are just extensions of the same narrowband speech coding techniques with a higher sampling rate. The most popular class of wideband codecs also comes with multirate adaptation by providing both low bitrate and high bitrate transmission. These types of codec ensure applicability to any underlying network condition and also eliminate the need for transcoding when the voice is routed from a high bandwidth network to a low bandwidth network. Next, the two recently developed wideband multirate codecs are described.

3.4.1 Adaptive MultiRate WideBand (AMR-WB)

The adaptive multirate wideband (AMR-WB) was jointly developed by VoiceAge and Nokia for the next-generation packet-switched wireless network. Currently, AMR-WB is standardized as ITU-T G.722.2 codec. AMR-WB extends the G.722 codec and is based on the ACELP coding technique. This codec is designed specifically for packet-switched networks by providing robustness to packet loss. The AMR-WB codec operation results in 20 ms frame size with 5 ms lookahead resulting in a total of 25 ms packetization delay. The codec provides excellent speech quality by having a higher sampling rate. The multiple bits rates supported by AMR-WB are follows: 6.60, 8.85, 12.65, 14.25, 15.85, 18.25, 19.85, 23.05 and 23.85 kb/s. The multiple bit rate modes can be changed dynamically to adapt to network congestion while ensuring good-quality reproduction. The lowest bit rate providing excellent speech quality in a clean environment is 12.65 kbit/s. Higher bit rates are useful in background noise conditions. AMR-WB is being standardized for deployment in future broadband wireless cellular network under the 3GPP standardization body.

There also exists an extended version of AMR-WB known as VMR-WB that supports both wideband and narrowband speech coding. The key feature of VMR-WB is that it is both a source-controlled and variable rate codec. The VMR-WB has four operation modes where the appropriate mode is selected based on the traffic condition and the desired QoS. The lower mode values $(0, 1, 2)$ are meant for operation in the CDMA2000 network supporting narrowband speech coding. In mode 3, the VMR-WB operates as AMD-WB.

3.4.2 Speex

Speex codec was designed specifically for VoIP over broadband connections. The design goals have been to make a codec that would allow both very good-quality speech and support multiple bit rates. Furthermore, Speex was designed to be robust to packet loss. Speex is based on CELP and supports ultrawideband (32 kHz), wideband (16 kHz sampling rate) in addition to narrowband (8 kHz sampling rate). Speex supports variable bit rate (VBR) encoding and Voice Activity Detection (VAD). Speex can provide a wide range of bit rates starting from 2 kb/s to 44 kb/s and thereby provides the ability to adapt to the available bandwidth. The Speex software design allows the complexity of the encoder to be controlled. By having codec software available as open-source and free from patent rights, Speex is being adopted by various VoIP softwares.

3.5 VoIP SOFTWARES

Recently, there have been various VoIP softwares or clients available for the PC that enables users to have voice communication over the Internet. The features of these softwares along with the supported codecs are discussed.

3.5.1 Linphone

Linphone is a good graphical SIP softphone (software-based phones), with support for several codecs. It works simply as a cellular phone. Linphone includes a large variety of codecs (G711-ulaw, G711-alaw, LPC10-15, GSM, SPEEX and iLBC). Due to supporting Speex codec, Linphone is able to provide high quality of voice even over slow Internet connections (such as 28 k modems). It also understands the Session Initiation Protocol (SIP) protocol which is a standardized protocol from the IETF. This guarantees compatibility with most SIP-compatible Web phones. It simply requires a soundcard to use with linphone. Other technical functionalities include DTMF (dial tones) support though RFC2833 and ENUM support (to use SIP numbers instead of SIP addresses). Linphone is free software, released under the General Public License. Linphone is well documented and includes a SIP test server called 'sipomatic' that automatically answers to calls by playing a prerecorded message.

3.5.2 SJphone

SJphone is a dual-standard VoIP softphone, which is well known for its good quality and interoperability. It allows the user to speak with any other softphones running on a PC, PDA, any stand-alone IP-phone, or using Internet Telephony Service Provider (ITSP) with any traditional wired or mobile phone. It supports both the SIP and H.323 standard sets, NAT traversal, and works with most major VoIP service providers such as Vonage, and IP-PBX and VOIP gateway vendors. SJphone is available for personal computers with Windows, MAC and Linux OS. SJphone supports GSM, iLBC, G.711 and G.729 codecs.

3.5.3 Skype

Skype is a proprietary peer-to-peer (P2P) Internet telephony (VoIP) network built using P2P techniques and competing against established open VoIP protocols such as SIP or H.323. The system has a reputation for working across different types of network connection (including firewalls and NAT). Skype users can speak to other Skype users for free, call traditional telephone numbers for a subscription fee (SkypeOut), receive calls from traditional phones for a fee (SkypeIn), and receive voicemail messages for a fee.

Skype uses wideband codecs which allows it to maintain reasonable call quality at an available bandwidth of 32 kb/s. Skype has a licensed VoiceEngine product, which is a comprehensive solution that includes all of GIPS codecs, as well as a jitter buffer, error concealment and echo cancellation technology. It is possible that Skype uses the GIPS ISAC codec. ISAC is a proprietary wideband variable rate codec, which can adapt its operating rate between 10 kbps and 32 kbps.

3.5.4 RAT

The Robust Audio Tool (RAT) is an open-source audio conferencing and streaming application that allows users to participate in audio conferences over the Internet. These can be between two participants directly, or between groups of participants on a common multicast group.

RAT requires no special features for point-to-point communication, just a network connection and a soundcard. For multiparty conferencing, RAT uses IP multicast and therefore all participants must reside on a multicast-capable network. RAT is based on IETF standards, using RTP above UDP/IP as its transport protocol, and conforming to the RTP profile for audio and video conferences with minimal control.

RAT features a range of different rate and quality codecs that includes G.711 PCM, Wide-Band ADPCM, G.726, GSM and LPC. RAT also provides performance improvement features such as receiver-based loss concealment to mask packet losses, and sender-based channel coding in the form of redundant audio transmission. It offers better sound quality relative to the network conditions than most audio tools available.

3.6 SUMMARY

Many years of research and development of speech coding techniques have resulted in a wide range of narrowband codecs differing in their complexity and bit rates. The narrowband codecs standardized by the ITU-T were designed for PSTN lines where minimizing the transmission rate was the primary objective. However, the situation has changed in the current VoIP deployments scenarios, where voice is delivered over packet-switch IP networks. Current IP networks provide higher bandwidth and can easily support wideband codecs, which suggests that lowering the codec bandwidth requirement is not crucial.

What is indeed important for the success of VoIP is the design of codecs that can adapt to the variation in the bandwidth, delay and be robust to the packet losses. The new class of VoIP codec starting with iLBC, AMR-WB is trying to address the new requirements of current and future VoIP deployments.

REFERENCES

1. ITU-T Recommendation G.711, Pulse Codec Modulation (PCM) of voice frequencies (1988).

2. ITU-T Recommendation G.726, Adaptive differential pulse codec modulation (1990).

3. Bradbury, J. Linear predictive coding (2000) (unpublished).

4. ITU-T Recommendation G.729, Coding of speech at 8 kbit/s using Conjugate-Structure Algebraic-Code-Excited Linear-Prediction (CS-ACELP) (1996).

5. ITU-T Recommendation G.728, Coding of speech at 16 Kbit/s using low-delay excited linear prediction (1992).

6. ETSI EN 301 704 V7.2.1 (2000–4), Adaptive Multi-Rate (AMR) speech transcoding (GSM 06.90 version 7.2.1 Release 1998).

)

CHAPTER 4

PERFORMANCE OF VOICE CODECS

One of the major challenges facing VoIP deployment is achieving and maintaining acceptable voice quality. The expectation is that the user-perceivable voice quality within a VoIP network has to match that of conventional circuit-based networks. This perceived voice quality is affected by many network-level factors including delay, jitter and packet loss.

The perceived quality depends on the specific codec used and its performance in handling the network-level factors. Therefore, it is critical to have appropriate network monitoring, management and planning techniques to ensure that VoIP traffic is handled properly and that VoIP quality is met. In addition, since call loads vary significantly with the sampling rates provided by different codec systems, such as G.711, G.729 and iLBC, the correct selection and usage of the VoIP codec system is crucial to have a good voice quality in the network under consideration. This chapter provides an understanding of the VoIP quality measures and provides methodologies for assessing and monitoring VoIP quality under different network conditions.

4.1 FACTORS AFFECTING VoIP QUALITY

VoIP quality is susceptible to the underlying network conditions (delay, jitter and packet loss). When a codec encodes a packet at the sender side, it contributes to these parameters, specifically to the delay. At the receiver side, when a codec decodes the packet(s) to reconstruct the voice content, the retrieved voice depends on these parameters that are introduced while the packets traverse the network.

4.1.1 Effects due to encoding

The delay caused by the codecs on the sender-side is reflected in the *packetization delay*. The amount of delay introduced depends on the codec speed. A codec intending to compress the signals so that a very low bandwidth is required will typically do more processing, whereas others can do with little processing.

Packetization delay due to coders in VoIP has three components:

- *Processing or Algorithmic Delay:* This is the time required for the codec to encode a single voice frame.

- *Look-ahead Delay:* This is the time required for a codec to examine part of the next frame while encoding the current frame. This helps to add information in the current frame that can be useful in decoding as set of frames as a whole. Most compression schemes employ the look-ahead technique.

- *Frame Delay:* This is the time required for the sending system to transmit one frame. This time depends on how voice samples are aggregated to create the packet in order to reduce the RTP-related overhead. For example, G.711 generates a voice packet every 20 ms by aggregating 160 voice samples.

This delay further depends on the processing capacity of the device on which the codec operates. On a device employing a fast CPU, even a processing-intensive codec will incur little packetization delay. This is the case where the voice terminal is a softphone running on a high-end computer. On the other hand, a device with limited CPU speed (such as mobile devices where battery consumption constraints mandate slower processors) will increase the packetization delay on any codec. Furthermore, packetization delay depends on the network interface card used in sending the VoIP packet into the network.

The lower layers, such as the MAC and the physical layers also perform frame aggregation before transmitting the packet. All the above factors contribute to the time delay incurred between receiving a voice sample and sending the corresponding VoIP packet/frame from the terminal.

4.1.2 Effects on the decoder

While the encoding portion of a codec introduces packetization delay in a packet, the decoding portion at the receiving end is affected not just by the packetization delay, but also by the additional delay introduced while packets traverse the network, the jitter in the delay and loss of any packets (resulting in lost voice samples).

- *Delay:* The delay introduced by various factors result in the packet reaching the decoder a while after the caller has spoken. From the perspective of decoding, all packets being delayed by the same amount of time does not impact the process. The packets that were created by the encoder at the rate of one every t time units, reach the decoder at the same rate of one packet every t units of time. By the design of the decoder, this does not impact the quality of reconstructed voice signals.

This does not mean that an inordinate amount of delay is acceptable for VoIP. The reason is that VoIP is an interactive application. It is necessary that the delay be minimized to have a good conversational quality. However, from the perspective of decoding, constant delay is not a factor.

- *Loss:* While traversing the Internet, the voice packets may be dropped at some router if it is congested. It can also happen in wireless networks due to bad channel conditions. This implies that the decoder will not receive all the packets created by the encoder. In this case, the decoder has to reconstruct the voice signal from a subset of the voice samples.

  Performances of different codecs vary in the presence of different degrees of loss. For example, codecs employing heavy PLC are likely to be able to reconstruct voice better than one with limited support for PLC.

- *Jitter:* The variation in delay faced by different packets is the jitter. To smooth out jitter, the receiver employs jitter buffer which tends artificially to delay the received packets, so that the decoder receives them at a constant rate (at which the encoder sent them). However, if the jitter is larger than a threshold, the jitter buffer is helpless and the decoder has to assume that the packet is lost (and continue without it). Such packets that are dropped due to a large jitter are termed as *jitter loss*.

Thus, the decoder is affected by the network parameters as it tries to reconstruct the voice at the receiver end. The impact of different parameters depends on the codec. We will discuss the performance of various codecs in detail later in this chapter.

4.1.3 Monitoring network conditions

In deploying VoIP applications, it is important to assess all these three characteristics (delay, jitter and packet loss) for expected situations in the network. Monitoring or measurements of delay, packet loss and jitter can help in a number of ways for VoIP service providers and equipment vendors. Statistics from monitored data can help in an understanding of the performance of a given codec and in determining the appropriate codec to be used with the VoIP software or in tuning the codec accordingly. In terms of network planning and operation, such statistics can help in correct configuration of traffic prioritization, as well as setting other QoS mechanisms.

Standard bandwidth monitoring techniques and traffic analyzers can provide basic information on network traffic and help to identify potential problems. The end systems can select different VoIP codecs and simulate different call loads during real-time tests. Different codecs, such as G.711 and G.729, provide different sampling rates that affect packet sizes and the bandwidth requirements.

4.2 VOICE QUALITY ASSESSMENT

It has always been challenging to define a good measure for VoIP quality. The main difficulty arises from the number of factors that affect voice quality and correlating these factors to the perceived quality. Typically, a measure should be defined in terms of a single metric that can be used to benchmark, analyze and tune various network and codec parameters. Therefore, not just correctness in terms of representing the VoIP quality, but the applicability of such a metric is also important. For example, if a metric is difficult to measure, the applicability of metric is reduced. The subjective MOS score (discussed in the next section) is typically measured by having real human listeners rate test suites of voice samples and taking a mean of the opinion scored based on the ratings of multiple listeners. In cases where the network parameters are tuned automatically based on real-time monitoring of observed VoIP quality, such a score is difficult to compute. However, there exist other metrics such as the R-factor (discussed later) that can directly create a score based on the real-time measurement of the network condition without human involvement. Indeed, R-factor has a wider applicability in terms of the network planning and configuration. However, the subjective MOS is more accurate than R-factor and has more applicability in design and testing of VoIP softwares by the equipment/software manufacturers.

With the wide range of codec choices, diverse network conditions to support and different deployment scenarios, it becomes necessary to understand the appropriate VoIP quality metrics and their measurement approach. In the following, we discuss the metrics that are relevant to the current VoIP deployment scenarios.

4.3   SUBJECTIVE MEASURES AND MOS SCORE

Subjective measurement of VoIP quality refers to a type of quality assessing methodology. Subjective measurement methodology results in a Mean Opinion Score (MOS) score that has the highest acceptability in terms of accuracy compared to other methods. Subjective measurements of the QoS are carried out by a group of people (test subjects) who listen to a test suite of voice phrases under different conditions. These tests are performed in special rooms, with background noises and other environment factors, that are kept under control for test executions. Some examples are: conversational opinion test, listening opinion test, interview and survey test.

The test methodology follows the following standard guidelines:

- number of listeners should be sufficient;

- listeners in normal hearing conditions;

- test suite consists of diverse phrases corresponding to different languages, genders, ages;

- controlled execution environment ensuring repeatability of experiments;

- proper choice of equipments used in the experiment (recorder, hearing aid etc).

Clearly, the methodology used in carrying out subjective measures is both expensive and time consuming.

There are three popular types of measurement methodology and the associated scores or rating. These three methodologies differ in terms of the listening test carried out to determine the quality rating. These methodologies are discussed next.

4.3.1   Absolute Category Rating (ACR)

The tests corresponding to Absolute Category Rating (ACR) result in a MOS. The tests are carried out by a listener hearing the sample at the received end without referring to the original sample. The exact guidelines for performing this test are specified in the ITU-T Recommendation P.800 [1]. The MOS represents the average value of the score assigned to the quality of the sample by the listeners. The MOS score is a numerical value between 1 and 5 where the score 5 refers to highest quality and 1 represents the lowest (or unacceptable) quality.

4.3.2 Degradation Category Rating (DCR)

The Degradation Category Rating (DCR) is used to rate high quality voice samples. This case, the ACR methodology is inappropriate to discover quality variations. This listening test is carried out using two samples: A and B. Sample A represents the reference sample with the reference quality, while sample B represents the degraded sample. Given these two samples, the listeners compare sample B against the corresponding sample A using a degradation scale. The comparison is expressed in terms of the Degraded Mean Opinion Score (Degraded MOS). The test is carried by having sample A followed by sample B separated by a silence without any particular ordering. The exact guidelines for test execution are specified in the ITU-T Recommendation P.800 [1]. The degraded MOS represents the average value based on the ratings presented by the listeners. Degraded MOS score is a numerical value between 1 and 5 where the score 5 refers to 'Degradation is imperceptible' while 1 represents 'Degradation is very annoying'.

4.3.3 Comparison Category Rating (CCR)

The experimental methodology underlying Comparison Category Rating (CCR) is similar to that of DCR except in the manner in which the test samples are used. Like DCR, CCR also uses samples A and B. In CCR, there is an exact ordering on how the samples are presented to the listener, i.e. sample A is always followed by sample B and the sequence is repeated. Therefore, in the case of DCR, the listener has an idea about which sample s/he is listening to. In contrast, with CCR, the listener has no clue as to the sample type (reference or degraded) to which s/he is listening. The exact guidelines for conducting the CCR test are specified in in ITU-T Recommendation P.800 Annex D [1]. The CCR score is defined in the range -3 to 3 where score 3 represents 'much better' and -3 score represents 'much worse'.

4.4 CONVERSATIONAL OPINION SCORE

The subjective scores described above are accurate in terms of understanding how the voice sample is reproduced and perceived at the receiver. However, this score does not always correlate to the user perception when the use is in an interactive mode. Such an interactive mode is common with VoIP. In the traditional telephony where the deployment was over the circuit switched network, the delay and jitter was not critical. For such a deployment, the above scores would suffice in terms of assessing the quality.

However, the situation is not the same with current VoIP deployment over the Internet and wireless medium where we have a packet-switched network with high delay and jitter. High and variable delay can easily degrade user experience in terms of communication comfort. This is the case even when the voice is perfectly reproduced at the receiver end.

In order to assess the interaction or conversational nature of VoIP, a more appropriate methodology is the conversational opinion test.

In this type of test, two subjects that perform the test are placed in separated and isolated rooms and undergo a conversation between themselves. Each conversation generates two opinions. Based on multiple such experiments, the mean score opinion is obtained, which is also referred to as the Mean Conversation-opinion Score. Conversational opinion test methodology and guidelines are defined by the ITU-T Recommendation P.800 Annex A.

4.5 E-MODEL

The E-Model defined by ITU G.107 [2, 3] is a computational model that can be used as a transmission planning tool for telecommunication systems. One novel feature of the E-Model is the assumption that the psychological effect of uncorrelated sources of impairment is additive. The assumption is based on empirical results in the field of psychophysical research, which relate physical stimulus magnitudes to perceptual magnitudes. The transmission rating or quality score is referred to as the R-factor (or R-score) and ranges from 0 to 100. The R-score is given by the following equation:

$$R = R_o - I_s - I_d - I_e + A.$$

Here, R_o represents the transmission rating of the basic signal-to-noise ratio. R_o factors in various types of noise that includes circuit noise, room noise at sender and receiver, sound of the speaker's own voice as heard in the speaker's telephone receiver (sidetones), and the noise generated by the device itself. The default value of R_o equals 93.2. This factor is the sum of all impairments which occur simultaneously with the voice transmission. Typically, a voice transmission is represented by an exceedingly loud voice signal, quantizing distortion due to various factors including Analog-to-Digital and Digital-to-Analog conversion, logarithmic PCM coding, ADPCM coding and a non-optimum talker sidetone.

The factor I_d represents this delay impairment, which is strongly effected by talker and listener echoes. If echoes are present, the delay can be noticed more easily. I_s is the signal-to-noise impairment associated with typical switched-circuit network paths. This term essentially covers the traditional PSTN-related quality impairments. I_e takes into account all impairments caused by more complicated and new equipment using the VoIP system. It is mainly used for predicting the coding distortion of low-rate speech codecs. I_e depends on the frame loss rate and the level of the effect depends on the type of coding and loss concealment technique used in the codec.

The last factor A is based on the knowledge that the quality of the telephone call is judged differently based on user expectation. A higher value of A generally refers to a lower expectation from users. For example, wireless, cellular and satellite connections are typically valued higher. Cellular phone users do not expect the same level of quality as in PSTN telephone calls. If the Internet access is cheap or even free, the expectation of VoIP users will be low, thus increasing the A and consequently the R-scores. Typical values of A range from 0 to 20.

The R-factor typically ranges from 0 to 100 and a score of more than 70 usually indicates that the VoIP stream has decent quality. For wideband codecs, the R-factor may increase above 100 and can reach 110 for an unimpaired connection. For wide-band codecs, the MOS score range is still 1–5 even though the R factor range is higher. This means that a narrow-band codec may have a MOS score of 4.3 and a wide-band codec may have a MOS score of 3.9, even though the wideband codec sounds much better. In general, one can correlate the R-score to the MOS using non-linear equation as shown in Figure 4.1.

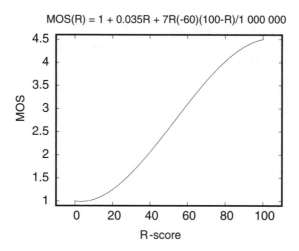

MOS(R) = 1 + 0.035R + 7R(-60)(100-R)/1 000 000

Figure 4.1 R-score to MOS score relation.

Among all of the factors in R-score computation, only I_d and I_e are typically considered variables in VoIP and require further understanding. Therefore, one can work with default values and use a simplified equation for the R-factor, as given below.

$$R = 94.2 - I_e - I_d.$$

Based on the above equation, the following discussion tries to explain how the R-score captures the sensitivity of voice quality to various impairments as they exist specifically in VoIP deployments.

4.5.1 Sensitivity to delay

The value of I_d captures the total mouth-to-ear delay which is composed of three components: codec delay d_{codec}, playout delay $d_{playout}$, and network delay $d_{network}$. Codec delay represents the algorithmic and packetization delay associated with the codec and varies from codec to codec. For example, the G.729a codec introduces a delay of 25 ms. Playout delay is the delay associated with the receiver-side buffer required to smooth out the jitter for the arriving packet streams. Network delay is the one-way transit delay across the IP transport network from one gateway to another. Thus, the total delay is given by

$$d = d_{codec} + d_{playout} + d_{network}.$$

The delay impairment, denoted by I_d, depends on the one-way mouth-to-ear delay experienced by the VoIP streams. This mouth-to-ear delay determines the interactivity of voice communication. Its impact on the voice quality depends on a critical time value of 177.3 ms, which is the total delay budget (one-way mouth-to-ear delay) for VoIP streams.

The effect of this total delay d is modeled as

$$I_d = 0.024d + 0.11(d - 177.3)\mathcal{H}(d - 177.3)$$

where $\mathcal{H}$ is an indicator function. $\mathcal{H}(x)$ is 0 if $x < 0$; otherwise it is 1.

Figure 4.2 shows the R-score as a function of delay and that the R-score drops relatively rapidly when the delay exceeds 177.3 ms. This observation is important when deploying VoIP to ensure that the delay of 177.3 ms is not exceeded.

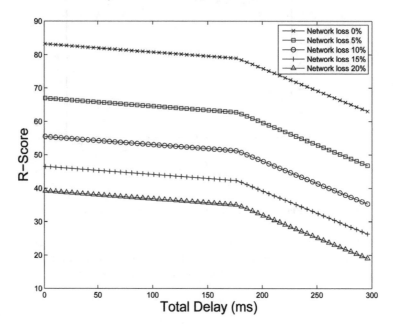

Figure 4.2   R-score as a function of delay.

4.6 SENSITIVITY TO LOSS

The VoIP call quality is also dependent on the loss impairment. Recall that I_e represents the effect of packet loss rate. I_e accounts for impairments caused by both the network's and the receiver's playout losses. Different codecs, with their unique encoding/decoding algorithms and packet loss concealment techniques, yield different values for I_e. Based on the E-model, I_e is related to the overall packet loss rate as

$$I_e = \gamma_1 + \gamma_2 \ln(1 + \gamma_3 e)$$

where γ_1 is a constant that determines the voice quality impairment caused by encoding and γ_2 and γ_3 describe the impact of loss on the perceived voice quality for a given codec. Note that e includes both network losses and playout buffer losses, which can be modeled as

$$e = e_{\text{network}} + (1 - e_{\text{network}})e_{\text{playout}}$$

where $e_{network}$ is the loss probability due to the loss in the network and $e_{playout}$ is the loss probability due to the playout loss at the receiver side. The values of the parameters $\gamma_1, \gamma_2, \gamma_3$ for the G.729a and G.711 codecs are shown in Table 4.1 below.

Table 4.1 Loss impairment parameters.

Codec	γ_1	γ_1	γ_1
G.729a	11	40	10
G.711	0	30	15

The dependence of the R-score on loss is shown in Figure 4.3. Notice that in the delay range (50–150 ms), the R-score depends mainly on loss and not on delay and this dependance is significant. When the loss is increased from 0% to 5%, the drop in the R-score is about 14. Typically, the delay in the Internet (unless congestion is experienced) is less than 100 ms. Consequently, for VoIP deployment over the Internet, loss is more crucial than delay. Similarly, the delay over the Wi-Fi link is less than 60–80 ms. In Wi-Fi as well, the quality is more dependent on the packet loss or packet corruption owing to wireless interference.

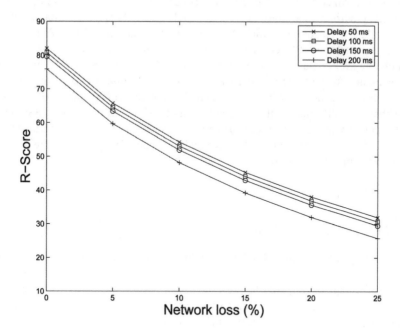

Figure 4.3 R-score as a function of loss.

The above discussion demonstrates that the R-score is an important metric that can easily provide a general sense of how well the network is equipped to handle VoIP without going through any cumbersome experiments like with the subjective scores.

4.7 PERCEPTUAL EVALUATION OF SPEECH QUALITY (PESQ)

Although the E-model is an excellent planning and decision tool, it cannot replace real measurements on the final network that can exactly match the transmitted waveform that is put in to the received waveform at the output and the corresponding human-level perception. In that regard, the challenge has always been to remove the human involvement part which is present in creating the subjective measures.

Perceptual Evaluation of Speech Quality (PESQ) provides such a measure and a methodology that makes quality assessment both inexpensive and less laborious. The advantage is that PESQ provides repeatability and can be used for performance benchmarks. Further, the ITU-T realized that subjective measures are not well equipped to capture a higher distortion level and longer delay and its variation as typically observed with VoIP. The development of PESQ was also prompted by the above requirement.

PESQ compares an original speech sample with the degraded version that was received. PESQ implements a cognitive model which emulates the psychoacoustics of human hearing. One novel feature of PESQ is its identification of transmission delays. First, PESQ adjusts the degraded version to be time aligned. Then, a psychoacoustic model assesses the distortion between the original and degraded samples. PESQ can identify both constant delay offset and variable delay jitter. One should note that PESQ can only be applied for distortions known before its development. These are coding distortions due to the effect of waveform codecs and CELP/hybrid codecs, transmission/packet losses, multiple transcoding, environmental noise and variable delay. Details about calculating PESQ are specified in ITU-T Recommendation P.862 and ITU-T Recommendation P.862.1.

Benchmark tests of PESQ have yielded an average correlation of 0.935 with the corresponding MOS values under these conditions. PESQ may have to be changed before it can be applied for low-rate vocoders (below 4 kbit/s), digital silence, dropped words or sentences, listener echo and wideband speech. The computational complexity of PESQ is high. Although PESQ (unlike the E-model) cannot be used in real-time, PESQ provides a more accurate voice quality score for networks, systems and devices compared to the E-model. This score is based on a subjective scale and can be mapped onto the results from thousands of subjective MOS results.

4.7.1 PESQ analysis for VoIP codecs

One disadvantage of the E-model is that it is defined only for standard codecs such as G.729 and G.711. With the recent popularity of new VoIP codecs such as iLBC, Skype codec and Speex, it is difficult to compare the performance of these new codecs with the E-model. However, one can use the PESQ model to compare the performance of various codecs (and VoIP softwares with proprietary codecs such as Skype).

In the following, we report PESQ scores for multiple codecs based on laboratory test results. To give a feel for the overall performance, the results are shown for four scenarios (low and high) loss rate and (low and high) jitter values in Figures 4.4–4.7. However, the observations discussed next are based on a more extensive set of results where the loss is varied from 0% to 14% and jitter is varied from 0 to 160 ms. Note that the large jitter variation covers the case of large delay as well. Table 4.2 shows the exact values for the extreme cases of loss and jitter.

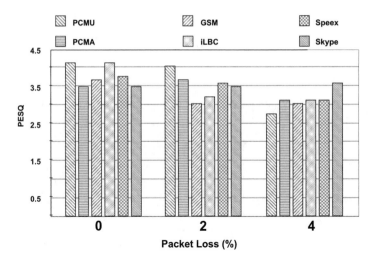

Figure 4.4 PESQ comparison with low packet loss rate (jitter = 0 ms).

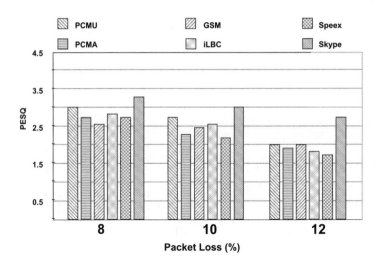

Figure 4.5 PESQ comparison with high packet loss rate (jitter = 0 ms).

The test results demonstrate the following characteristics. The test result for the case where there is no delay in the networks shows that PCMU and PCMA have the highest PESQ score of above 4.0 and GSM codec has the lowest score of 3.5.

As packet loss is increased, the codec for Skype shows relatively more robustness compared with other codecs. For example, all codecs show almost the same voice quality of over 2% packet loss at the delay variation of 40 ms, but increased packet loss shows a greater effect on all codecs except Skype. Furthermore, as the delay variation increases, Skype's codec offers better robustness to the packet loss. For instance, Skype has a PESQ score of 2.6 while other codecs have PESQ scores of around 1.7 with jitter 60 ms and 20% packet loss. The main reason for resilience to packet loss for the Skype codec is

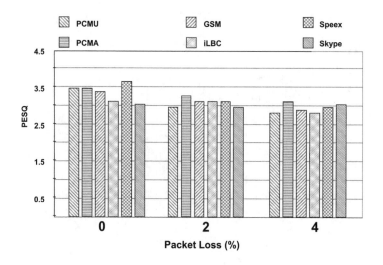

Figure 4.6   PESQ comparison with low packet loss rate (jitter $= 40$ ms).

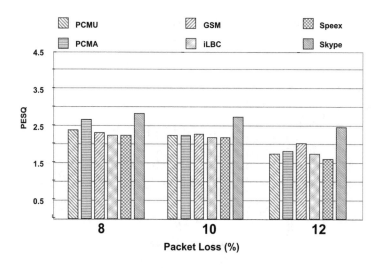

Figure 4.7   PESQ comparison with high packet loss rate (jitter $= 40$ ms).

that it actually uses two different codecs, iLBC and iSAC and switches usage of them depending on the network resources and host's hardware resources. Although the details of the algorithm are not publicly available, the test result clearly shows that the Skype algorithm provides better performance than others. In fact, for the case where we have a high jitter of 150 ms, while other codes have PESQ scores below 2.0, the Skype codec has relatively high scores for all the cases of packet loss below 20%.

Since each codec has a different packet size and bandwidth for network traffic, packet loss and delay may have different impacts on their performance. PCMU and PCMA codecs have the best performance when network delay is low and packet drop is at the minimum rate. Their PESQ values decrease dramatically when the jitter is more than 80 ms.

Table 4.2   PESQ comparison for various codecs.

Scenarios	PCMU	PCMA	GSM	iLBC	Skype	Speex
Jitter: 0 ms Loss: 0%	4.130	4.144	3.834	3.858	3.515	4.139
Jitter: 0 ms Loss: 20%	1.939	1.946	2.013	1.908	2.922	1.749
Jitter: 150 ms Loss: 0%	2.120	1.915	1.858	2.087	2.026	1.692
Jitter: 150 ms Loss: 20%	1.315	1.595	1.796	1.248	2.083	0.929

The GSM codec does not show good PESQ scores regardless of jitter and packet loss condition. It uses 13 Kbps of network bandwidth and its performance is relatively better than PCMU and PCMA whose network bandwidth usage is 64 Kbps.

iLBC does not show impressive performance when compared with other codecs but the test result shows that it is more resilient to packet loss than PCMU and PCMA. Speex produces good PESQ scores with low jitter and is resilient to considerable packet loss. Since Speex codec was designed to operate in a VoIP network, it is robust to packet loss of up to 10%, but the PESQ score is not better than others when packet loss is more than 10% or when the jitter is more than 150 ms.

4.7.2 Cross correlation

Cross correlation is a standard method of estimating the degree to which two signals are correlated. Cross correlation coefficient values (ranging from -1 to 1) indicate to what degree two signals are correlated. Therefore, the value of correlation provides a sense of the level of distortion of the received signal in comparison to the original. If its maximum value is close to 0 or -1, it means that the received signal has been severely distorted due to jitter or packet losses. If it is close to 1, the degraded signal still maintains the originality of the reference signal. Cross correlation values can also capture the effect of jitter.

However, sound quality is not completely dependent on cross correlation. Thus, if one signal has a higher cross correlation with the original signal compared to that of another signal with the original, it does not necessarily imply that the received voice quality is better in the first case. Yet, cross-correlation can be useful in studying the relative effects of network parameters on codecs, particularly for comparing proprietary codecs such as those of Skype.

The cross correlation coefficient values for different codecs with different network parameter settings are shown in Figures 4.8–4.10. The results show the coefficients when loss and jitter are introduced. For the case jitter 0 ms, 0% loss, the maximum score of cross correlation for Skype is the highest due to the nature of the wideband codec utilized by Skype.

4.8 TOOLS FOR LAB TESTBED SETUP

There exist various commercial tools for measuring the PESQ. In this section, we present a simple and affordable way of generating PESQ in a laboratory environment. This methodology is appropriate for conducting preliminary experiments on quality assessment. The experimental testbed has the following parts:

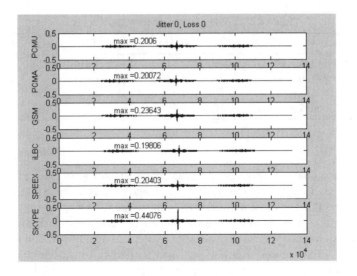

Figure 4.8 Cross correlation coefficient with zero loss and jitter.

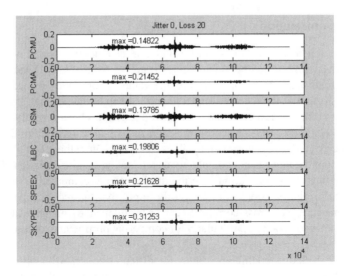

Figure 4.9 Cross correlation coefficient with loss.

- a network emulator that can emulate various loss/delay/jitter effects in a controlled way;

- voice input/output tools;

- a recording tool.

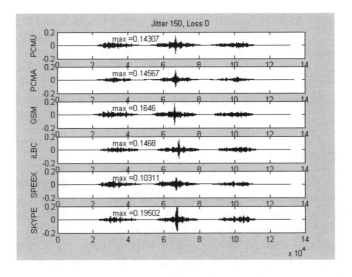

Figure 4.10   Cross correlation coefficients with jitter.

4.8.1   Network emulator

The NIST Net [4] network emulator is a general-purpose tool for emulating performance dynamics in IP networks. The tool is designed to allow controlled, reproducible experiments with network-sensitive applications in a simple laboratory setting. By operating at the IP level, NIST Net can emulate the critical end-to-end performance characteristics imposed by various wide-area network situations (e.g. congestion loss) or by various underlying subnetwork technologies (e.g. asymmetric bandwidth situations of xDSL and cable modems).

NIST Net is implemented as a kernel module extension to the Linux operating system and an X Window system-based user interface application. The tool allows an inexpensive PC-based router to emulate numerous complex performance scenarios, including: tunable packet delay distributions, congestion and background loss, bandwidth limitation, and packet reordering/duplication. The X interface allows the user to select and monitor specific traffic streams passing through the router and to apply selected performance effects to the IP packets of the stream. In addition to its interactive interface, NIST Net can be driven by traces produced from measurements of actual network conditions. NIST Net also provides support for user-defined packet handlers to be added to the system. Examples of the use of such packet handlers include: time stamping/data collection, interception and diversion of selected flows and generation of protocol responses from emulated clients.

4.9   VOICE INPUT/OUTPUT TOOLS

JACK [5] is a low-latency audio server, written for POSIX-conformant operating systems such as GNU/Linux and Apple's OS X. It can connect a number of different applications to an audio device and can let them share audio as well. Its clients can run in their own processes (as normal applications), or they can run within the JACK server (as a 'plugin').

JACK was designed from the ground up for professional audio work, and its design focuses on two key areas – the synchronous execution of all clients, and low latency operation. In a basic testing configuration as shown in Figure 4.11, the audio server plays audio file in a WAV form and its output is redirected to an audio router such as JACK. VoIP applications use the output of the audio router (JACK) as an input and send them to the receiver. The received audio is forwarded to an audio recorder such as *Ardour* and *Skype-rec* through the audio routers.

4.9.1 Recording tools

Ardour [6] is a multitrack recording and editing system for high-quality digital audio. Ardour supports audio processing plugins, parameter control automation, sophisticated panning control and many advanced editing procedures. Recognized synchronization protocols include MIDI time code (MTC), a means of encoding SMPTE time code to MIDI; MIDI machine control (MMC), a set of MIDI messages for controlling transport features of hardware mixers and recorders; and JACK, a low-latency audio server and application transport control interface for Linux and Mac OS X.

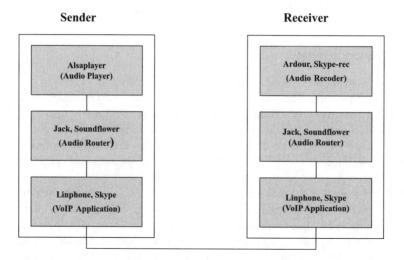

Figure 4.11   Configuration for a basic testbed.

4.9.2 Experiment configurations

In a basic configuration, a free VoIP software such as *Linphone* can be used which supports various codecs. The *Linphone* is used with the pair of JACK and *Ardour* to send, receive and record speech. The traffic generated is forwarded using NIST. The general configuration is shown in Figure 4.11. In this configuration, the pair of JACK and *Ardour* can be used with *Linphone* on Linux while the pair of *Soundflower* and *Skype-rec* is used with Skype. The different configuration for Skype is because JACK does not work with Skype. With the above configuration, NIST Net emulator can be used to control network attributes such as packet loss and jitter.

4.10 SUMMARY

The perceived quality of VoIP depends on various parameters among which the critical ones are loss and jitter. Active monitoring and management of voice quality in a VoIP environment is crucial to help identify and eliminate the effect of network parameters on quality degradation. Since voice quality is vital to the success of a VoIP phone system, monitoring around the clock pro-actively and identifying service faults allows providers the ability to improve their performance before end-users detect a problem. For such real-time monitoring, E-model has significant advantages. On the other hand, a more off-line analysis of VoIP quality using PESQ measures can be effective in performance comparison and selection of wide choices of VoIP codecs available. The quality assessment can also be leveraged for designing sophisticated algorithms for delivering VoIP, such as traffic scheduling algorithm, call admission algorithms and VoIP traffic routing algorithms.

REFERENCES

1. ITU-T Recommendation P.800, Methods for subjective determination of transmission quality, (1996).

2. ITU-T Recommendation G.107, The E-Model: A Computational Model for Use in Transmission Planning, (1998).

3. Cole, R.G. and Rosenbluth, J.H. Voice over IP Performance Monitoring, *Computer Comm. Review* (2001) **31**(2).

4. NistNet: http://snad.ncsl.nist.gov/nistnet/

5. Jack: http://jackit.sourceforge.net/

6. Ardour: http://www.ardour.org/

CHAPTER 5

VoIP PROTOCOLS

The ease and cost of deployment along with the supreme quality that VoIP can provide are the main drivers behind its tremendous growth. To make VoIP work, there are several components that have been designed and implemented. One such set of components comprises the communication protocols that enabled the VoIP-capable endpoints to communicate. These protocols vary from those that provide the simplest form of connectivity to those that are very sophisticated in optimizing the VoIP experience. The tasks that these protocols perform include resource discovery, connectivity establishment and call maintenance.

This chapters describes a small subset of such protocols. The intent is to provide the reader with a flavor of what are the motivating factors in designing different protocols. This chapter is not meant to provide an exhaustive treaty on any single protocol. The focus is on the underlying operations and the concepts therein.

5.1 INTRODUCTION

Protocols define the mechanisms that participants in a communication session must follow, depending on the task that they are trying to accomplish. In general, they describe the formats of the messages exchanged between the entities involved in a communication session, the different types of message possible along with their meanings in various contexts, the sequence of operations that need to be performed across entities in order to execute some task, and exception handling when some task expected to be executed does not materialize.

In more abstract terms, the protocols essentially define the state transition machines of each entity involved in a communication session. Receipt and transmission of messages along with expiry of timers typically result in transition of states at a participant entity. A sequence of message exchanges aims to drive the state machine to a desirable state. The definition of what constitutes a desirable state varies depending on the objective of the application. For example, for TCP the objective is reliable data transfer. The state transition machine traverses through various states including the three-way handshake for connection establishment, sending data and waiting for acknowledgements, retransmitting in case the acknowledgement is not received in a reasonable amount of time and repeating this process until all the data to be transferred has been acknowledged.

In the context of VoIP, the protocols are meant to establish the end-to-end voice call. Broadly speaking, from the perspective of VoIP there are two types of protocol:

- *Signaling Protocols:* These protocols are used to perform the auxiliary function related to setting up and maintaining a call. These tasks include:

    1. *Callee Location:* This involves finding out the current location (in the network) of a callee when someone intends to communicate with him/her. The signaling protocol has to communicate with the appropriate authority to find the user.

    2. *Availability Determination:* The protocol must decide whether the called party is available and if not, then where the call is to be redirected. For example, a busy callee might have set all calls to be redirected to voicemail. The signaling protocol needs to figure this out.

    3. *Session Parameter Negotiation:* The caller and callee must agree upon the parameters used in the media transfer in order to communicate. These parameters include the type of media, the codec to be used, etc. The signaling protocol will help both the parties agree upon the parameter set.

    4. *Session Modification:* While a call is ongoing, the parties may decide to change the content of the media transfer. For example, the codec might be changed to improve the quality if the end-to-end path has better bandwidth. Such an intent has to be signaled to the other party.

    5. *Session Termination:* When the parties involved in the communication decide to end the call, they need to signal properly to each other and possibly to other entities that the call is completed.

- *Media Transport Protocols:* These protocols are used for the actual media transport in a call. The voice packets are carried between the parties using these protocols. Thus, encoding/decoding, packetization and transport of the VoIP packets are done by these protocols.

We shall look at some of the protocols that help in providing these functionalities in the VoIP world.

5.2   SIGNALING PROTOCOLS

The two most prominent protocol suites for VoIP are ITU's H.323 and IETF's Session Initiation Protocol (SIP). Both support a very rich feature set for VoIP and have a significant deployment. We will look mainly at these two protocols.

5.2.1   Session Initiation Protocol (SIP)

SIP is a signaling protocol used for establishing sessions over the Internet. A session could be as simple as a two-way telephone call or it could be as rich as a full-fledged multiparty multimedia conference. In fact, any application which has a notion of a session can use SIP for the session setup, e.g. conferencing, telephony, presence, events notification, network games and instant messaging. The flexibility and capabilities of SIP make it an attractive choice for integrating different types of communication application into the existing data networks. Using SIP, telephony becomes just another Web application and integrates easily into other Internet services allowing providers to build converged voice and multimedia services. Hence, SIP is fast becoming a major player in the signaling protocol arena with significant deployment in the VoIP application domain.

5.2.1.1   Architecture overview   SIP is a simple protocol that serves purely as a mechanism to establish sessions. Its operation is independent of the underlying network transport protocol and indifferent to the content/type of the session being established (i.e. video, telephony, messaging, etc.). SIP simply defines how one or more participant devices can initiate, modify and terminate sessions. This simplicity means that SIP is flexible and can be used in different deployment scenarios.

SIP is a layered protocol implying that the protocol behavior is defined in terms of independent and loosely coupled processing layers. This makes the description of the protocol conceptually simple. The lowest layer of SIP specifies the physical syntax and encoding. This is described in the Backus-Naur Form in the corresponding RFC 3261 [1]. The next layer is called the transport layer, which specifies how clients and servers send and receive messages (requests and responses). The third layer is called the transaction layer. Any task that a client wishes to perform is executed as a sequence of transactions. This layer is responsible for matching responses to requests, and handles the application-layer retransmissions for reliability. Above the transaction layer is the Transaction User (TU) which represents almost all the SIP entities (except stateless proxies). When any TU has to send a request, it creates a client transaction instance and passes it to the request along with the parameters identifying the destination.

We note that SIP itself is not sufficient to create new applications, and different protocols need to come together for this purpose. For example, in order to provide telephony services, media transport protocols such as RTP, user authentication protocol such as RADIUS, directory service providing protocol such as LDAP, and quality of service protocols such as RSVP may be used. Here, we will only give a brief overview of SIP discussing primarily the underlying concepts rather than completely elaborating on all possible functionalities of SIP. The reader is referred to the corresponding IETF RFC 3261 [1] for more details.

SIP is a request–response protocol where each message is either a request or a response to a request. It closely resembles two other Internet protocols, HTTP and SMTP (the protocols that power the Web and e-mail respectively). Its commands are in clear-text and human readable similar to HTTP with many of the response codes being reused

Table 5.1 Basic request methods and response codes used in SIP.

Basic methods	Response code prefix	Purpose
REGISTER		Update Registrar's database with user's location
INVITE		Call Initiation
ACK		Acknowledgement for INVITE/BYE
CANCEL		Stop the execution of another command (such as INVITE)
OPTIONS		Find out a server's capability
BYE		Terminate a call
	1xx	Provisional – indicates that request is being processed
	2xx	Success – request was successfully understood and accepted
	3xx	Redirection – further action required to complete the request
	4xx	Client Error – bad syntax in request or unable to complete the request at this server
	5xx	Server Error – server failed to execute a seemingly valid request
	6xx	Global Failure – request cannot be completed at any server

(e.g. 200 – OK, 404 – Destination Not Found). A summary of basic SIP methods and response code prefixes is given in Table 5.1. The SIP addresses (called SIP Uniform Resource Identifiers) are similar to e-mail addresses (e.g. sip:foo@bar.com) and in practice, the SIP addresses are e-mail IDs of the users. SIP can also support other types of address such as conventional telephone and fax numbers.

5.2.1.2 SIP components

The SIP protocol is very feature rich and has several components. The key components of a SIP based system are:

- *User Agent (UA):* The user agent is an entity which forms an endpoint of a SIP session. In a client role, it initiates a SIP session and in a server role, it listens to client requests. UA can be implemented in hardware or software.

- *SIP Registrar:* The SIP Registrar is the entity with which a user registers its identity and location. The registrar maintains a location server to keep track of the current position of the user. This enables SIP to be used in presence-based applications where the current location and status of a user determines the actions needed to be taken with an incoming call.

- *SIP Proxy Server:* The SIP proxy server relays the incoming/outgoing signaling messages to the appropriate entities. On receiving a call request, it contacts the registrar to locate the target UA, forwards the request to the UA, and relays the response from the UA to the contacting client.

  In the simplest operation, the Proxy does not remain in the path of the call once the connection is set up between endpoints. However, it has the capability to remain in the media path if so required. This may be the case when, for example, the call needs to be billed by the duration.

Note that the aforementioned list of tasks for each component is not an exhaustive list. In order to provide the rich feature set, these entities may perform other tasks as required.

5.2.1.3 SIP operation We now go over a basic communication session using SIP where a user Alice (URI – sip:alice@domain1.com) wants to establish a SIP session with user Bob (URI – sip:bob@domain2.com). A portion of the process is illustrated in Figure 5.1.

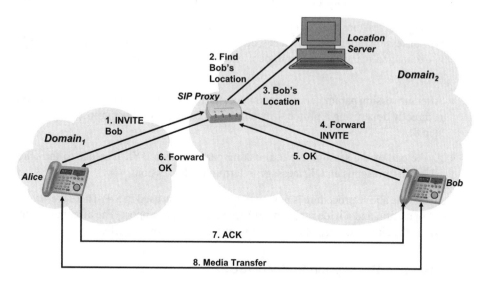

Figure 5.1 A basic SIP call setup operation.

- Alice and Bob register with the SIP registrars of their respective domains by sending a REGISTER message. The REGISTER message is used to update the location database at the registrar to bind their URIs with their respective IP addresses. This step is not repeated each time a user wants to communicate and is orthogonal to the call setup process.

- Alice's UA issues a DNS lookup to find the SIP proxy for domain domain2.com. This is done using the SRV message type in DNS request. The DNS reply contains the IP address of the SIP proxy for domain2 – proxy.domain2.com. Alice's UA can now contact this IP address to contact Bob.

- Alice's UA sends an INVITE message addressed to sip:bob@domain2.com to the SIP proxy server proxy.domain2.com (Figure 5.1, step 1).

- On receiving the INVITE message, proxy.domain2.com contacts the location server of domain2 to find the current location associated with sip:bob@domain2.com (Figure 5.1, step 2).

- The location server responds to the request sent by proxy.domain2.com and returns the current address of Bob (Figure 5.1, step 3).

- Proxy.domain2.com forwards the INVITE message to the UA at Bob's current IP address (Figure 5.1, step 4).

- Bob's UA receives a 'ring' (software or hardware ring depending on the type of UA).

- When Bob 'picks up' the phone, an OK status is returned to the proxy which forwards it to Alice (Figure 5.1, steps 5 and 6).

- Alice sends an ACK directly to Bob since she now knows his location (and there is no need of proxy to locate him) (Figure 5.1, step 7).

- During this session, the INVITE and OK message also contain information about the type of session (audio/video, codecs, etc.) Alice and Bob can handle. This information is exchanged using the SDP protocol whose content is carried inside the INVITE message.

- After the session parameters are negotiated, the media transfer between Alice and Bob is directly done using a protocol such as RTP designed for this purpose (Figure 5.1, step 8).

- At the end of the session, the terminating party sends a BYE message to the other which in turn sends an OK message to terminate the session.

Note that the above procedure is not the only one which is used in SIP. There are several modes of operation and a wide range of features that SIP can support. The aforementioned procedure is the simplest form of connection setup and tear-down in SIP.

5.2.2 Session Description Protocol (SDP)

As the name suggests, Session Description Protocol (SDP) is used to describe the parameters of a session. The parties involved in a communication session can have different capabilities. When initiating a session between these devices, it is necessary for the two parties to agree upon a common framework for information transfer. For example, if both parties have a capability to encode/decode a voice call from a multitude of codecs, they need to agree upon which codec they will use.

SDP provides a simple way of negotiating such parameters. While defined as an independent protocol, it is the session description framework of choice for SIP and is usually associated with it. SDP forms part of the SIP INVITE message containing information that allows the participants to decide upon the format of communication. SDP has been designed for general session description and is not limited to a voice call. For example, it can be used to describe a conference.

Among other fields, SDP includes information about the media that comprises the session (in form of the type of media, the transport protocol used, the media format), and the information to receive the media (such as the IP address and port number). Using this information the parties can agree upon the baseline for their media transfer. More details about SDP can be found in the corresponding IETF RFC 2327 [2].

5.2.3 H.323

H.323 is an ITU proposed standard to support media transport. It was designed specifically for all forms of media (audio and video) and data communications over IP-based networks (including Internet). Hence it is applicable for VoIP as well. In fact, it is widely used in the present VoIP deployments.

Fundamentally, H.323 is an umbrella standard for all forms of multimedia communication over networks that do not have support for QoS guarantees. At its core, it is designed to handle the fluctuations in the packet transmission latency by the network. Since most of the existing networking technologies do not support QoS guarantees and hence have variable packet delays, H.323 is applicable across all of them. H.323 encompasses various technologies, end-devices, along with conferencing with multiple participants. Furthermore, it provides a mechanism to interconnect endpoints residing in networks with different technologies.

5.2.3.1 *H.323 architecture overview*   Like SIP, H.323 is designed to enable a peer-to-peer connection between intelligent endpoints. The H.323 components and protocols act together to facilitate the endpoints to communicate. The protocol suite is very powerful and can provide a range of features that are useful in multimedia communications. At the same time, the architecture is flexible enough to allow different devices operating on different network technologies to communicate without worrying about the capabilities of the other endpoint. H.323 is designed to work independently of the underlying network technology. Its scope does not mandate any particular transport layer connecting the endpoints.

The H.323 messages are defined using Abstract Syntax Notation number One (ASN.1) which is a standardized formal language to define messages in an abstract form. The messages are encoded for transmission in a compact binary format resulting in efficient utilization of bandwidth.

From the perspective of scalability, the key component that H.323 has to manage is the gatekeeper (described next). H.323 has the capability to balance load across a multitude of gatekeepers (described next) when faced with large call volumes. Furthermore, it can operate in the direct call mode where the gatekeeper handles only a single message for address resolution and can operate in a stateless manner.

Like SIP, H.323 also separates the control plane from the data plane. It uses RTP as the media transport protocol. As its underlying transport protocol, H.323 can use both UDP and TCP. In practice, most of the signaling is done over TCP. Furthermore, H.323 is independent of the codec used and can work with any codec.

5.2.3.2 *H.323 components*   There are four major components defined in H.323. These are:

- *Terminal:* In H.323, *terminals* denote the endpoints that can provide real-time, two-way communications. At the very least, the communication capability of a terminal should include support for voice. It may or may not support video and data communication.

  Along with the real-time two-way communication requirement, terminals are required to support other standards that fall under the H.323 umbrella. These include support for Q.931 for call signaling, for H.245 to negotiate parameters and capabilities, and for the Registration/Admission/Status(RAS) component to communicate with the gatekeeper. Lastly, the terminal must also support media transport protocol such as RTP as these protocols carry the actual media.

  Depending on the other features implemented in the terminal, it can have a gamut of codecs allowing it to connect to devices with varying degrees of capability, data conferencing capability if support for T.120 protocols is built in, and MCU capabilities.

- *Gateway:* In H.323, a gateway is a component that acts as the bridge between devices residing on different types of network. A gateway provides translation facilities to such devices. It can perform various types of translation such as that between codecs and transmission formats. The most common-use scenario of a gateway is to connect an H.323 terminal to a PSTN terminal.

Since the purpose of gateway is to facilitate communication across disparate networks, they are not a mandatory part of the H.323 network. If we know that all the communication between the endpoints is going to remain on the same network segment, there is no need for a gateway.

Terminals communicate with gateways using the H.245 and Q.931 protocols.

- *Gatekeeper:* A *gatekeeper* is the core component of an H.323 network. The terminals in the network register with the gatekeeper. Collectively, all the devices (terminals, gateways, MCUs) managed by a single gatekeeper form its *zone*. The gatekeeper provides call control services to all registered endpoints in its zone. Also, it controls the access to network resources inside its zone. This makes it a key component in the H.323 network.

A gatekeeper performs the address translation from the H.323 alias of a device to its network address (e.g. IP address). Like SIP Registrar, it maintains a table mapping the alias to the address and the table is updated by the terminal registration messages. Typically, the alias is the e-mail address or the E.164 number. Furthermore, it performs the critical task of bandwidth management inside the network. This it achieves by limiting the number of calls active in the system. By controlling the number of calls below a threshold, a gatekeeper is able to control the amount of collective network bandwidth in use. Also, the calls can be denied based on the access policy of the network.

Similarly to the SIP Proxy, the gatekeeper can keep out of the media path. However, if need be, it can remain in the media path. As mentioned earlier, this may be useful in certain cases such as billing for the call by the usage. Furthermore, call rerouting can be done by the gatekeeper to balance the translation load among multiple gateways.

Like a gateway, a gatekeeper is an optional component in an H.323 system. Without a gateway, there is no bandwidth management functionality in the network. However, if the network does have a gateway, the terminals should become part of its controlled zone.

- *Multipoint Control Unit (MCU)*

The H.323 specification supports conferencing between multiple endpoints using the Multipoint Control Unit (MCU). An MCU has two components: Multipoint Controller (MC) and Multipoint Processors (MP). Of these, MC is mandatory and while MP is optional, there can be multiple MPs in the system.

The MC forms the core component of the MCU which controls the conference specifics. These include handling the H.245 negotiations to determine the shared capabilities among all parties. The MC handles H.245 negotiations between all terminals to determine common capabilities for audio and video processing. Also, it can decide which of the media streams should be multicast.

The MP deals with the media streams themselves. It handles the mixing and switching of streams.

5.2.3.3 *H.323 protocols* H.323 is a protocol suite which defines individual protocols for different tasks. We look at a small sample of these protocols that are useful in basic call setup in H.323. These three protocols are the main system control protocols in H.323.

- *RAS:* As the name implies this is the protocol that is used by the terminals to communicate with the gatekeeper. All signaling with the gatekeeper is done using RAS. An H.323 system does not use RAS if there is no gatekeeper in the system. RAS is used to register with the gatekeeper, requesting admission for a call, bandwidth control, name resolution, and status reports.

- *Q.931:* In the H.323 suite, Q.931 protocol is used to establish and tear-down a connection between two terminals. Originally, Q.931 was the connection control protocol for ISDN and was adapted from there for H.323 as its call signaling protocol. This use in H.323 also involves the use of H.323-specific data in the Q.931 UUIE (User-User Information Element) element of Q.931. It contains messages that allow creation, suspension, resumption, and termination of calls along with mechanism to report the current call state.

- *H.245:* H.245 is the call control protocol under the H.323 umbrella. It is used to open and close logical channels between endpoints and to negotiate capabilities in order to agree upon the call parameters. There is one H.245 control channel for each call. H.245 provides a reliable communication capability. After the call is set up (using Q.931), H.245 is used to determine the call parameters by its capability exchange mechanism where each terminal can inform the other of the type of media it is capable of transmitting and receiving. Even after a call has been established, H.245 can be used to change the parameters of the call. The H.245 control channel is always open.

5.2.3.4 *H.323 operation* A simple communication session using H.323 is described below. There are two terminals in this case; one belonging to Alice and the other to Bob. Alice wishes to communicate with Bob. A portion of the sequence in a typical H.323 setting is shown in Figure 5.2. The communication takes place as follows:

- Before initiating any calls, Alice and Bob will register with their respective registrars using the Registration Request (RRQ) message of RAS. The gatekeeper registers their location and conveys them a successful registration by sending the Registration Confirmed (RCF) message of RAS.

- Alice sends an Admission Request (ARQ) message of RAS to the $Zone_1$ gatekeeper indicating that she wishes to talk to Bob in $Zone_2$ (Figure 5.2, step 1).

- The gatekeeper tries to find out the location of Bob by sending the Location Request (LRQ) message of RAS to the $Zone_2$ gatekeeper. (Figure 5.2, step 2).

- The gatekeeper of $Zone_2$ confirms the location of Bob by sending back the Location Confirm (LCF) message to the gatekeeper of $Zone_1$ (Figure 5.2, step 3).

- The gatekeeper of $Zone_1$ admits Alice's call request and responds back to Alice with the Admission Confirm (ACF) message of RAS (Figure 5.2, step 4).

- Alice sends a Q.931 SETUP message to Bob to probe whether a call can be initiated (Figure 5.2, step 5).

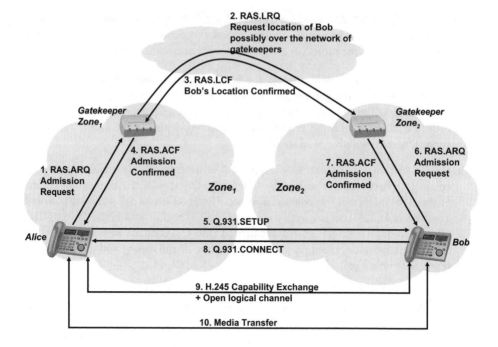

Figure 5.2 A basic SIP call setup operation.

- Now Bob asks its zone's gatekeeper permission for call admission by sending a RAS ARQ message (Figure 5.2, step 6).

- $Zone_2$'s gatekeeper admits Bob's call and responds in the affirmative by sending the RAS ACF message (Figure 5.2, step 7).

- Now that the call has been admitted locally, Bob sends back a Q.931 CONNECT message to accept the call (Figure 5.2, step 8).

- Next, the two parties negotiate the call parameters such as the codec to be used using the H.245 messages. Also, a logical H.245 control channel is established to exchange any relevant information during the call (Figure 5.2, step 9).

- Once all the setup formalities have been completed, the two establish an RTP channel for the actual media transfer (Figure 5.2, step 10).

Note that this is one specific scenario of communication using H.323. In practice there are simpler as well as more complex scenarios that H.323 can handle.

5.2.4 Media Gateway Control Protocol (MGCP)

While SIP and H.323 are end-to-end call control protocols, Media Gateway Control Protocol (MGCP) is used to control the telephone gateways from an external control component called Media Gateway Controller (MGC) or a call agent. As discussed in the context of H.323, a gateway is an entity which acts as a translator between disparate endpoints, e.g.

converting voice signals on the PSTN side to the digital packet format on the IP side and vice versa.

5.2.4.1 *Components*   There are two main components in the MGCP terminology.

- *MGC:* In the MGCP system, the MGC is the master. It contains the intelligence necessary to control the operation of the slave media gateways.

- *Media Gateway (MG):* The MGs are the slaves in the MGCP. They take the commands from the MGC and execute them accordingly. As gateways, their task is to translate codecs, signals, protocols, etc., depending on the types of network they mediate.

5.2.4.2 *Architecture overview*   MGCP is designed as an internal protocol to coordinate the operations of components inside a distributed system that acts as a single VoIP gateway with respect to the external world. This distributed system is organized in a master–slave configuration where the master controls the operation of the slaves. The slaves are the MGs and the master is the MGC.

MGC uses MGCP to inform the MGs under its control of all information regarding the call including the call parameters, capabilities and routing information. In order to inform the gateway of the call parameters, MGCP uses the convention followed by SDP. The commands issued by MGC are in text format. MGCP does not provide any specifics of how such information is obtained by the MGC. In general, the MGC can obtain this information by communicating with the other endpoint of the call using any existing protocol including SIP and H.323.

The connection model in MGCP consists of endpoints and connections. The endpoints can be physical endpoints such as the interfaces on the gateways. A virtual endpoint can be an audio source in an audio server. Essentially, the endpoints are the sources and sinks for all data. Connections associate endpoints. Connections may be point-to-point where two endpoints associate to transmit data. In a multipoint connection, multiple endpoints can connect. Connections are independent of the underlying networks. A call comprises a group of connections. Each call is set up by one or more MGCs. In essence, MGC identifies the set of connections that form a call and instructs the MGs to create those connections and whether they are to physical or to virtual endpoints.

5.2.4.3 *MGCP operation*   MGCP does not define how the MGCs themselves communicate and synchronize with each other. Thus, the MGC can possibly communicate using other protocols such as SIP/H.323. A simple setting of the use of MGCP is shown in Figure 5.3. Alice wishes to communicate with Bob and both their phones are connected to MGs which are under the control of an MGC.

- Alice picks up the phone and dials Bob's number (Figure 5.3, step 1).

- The media gateway for Alice (GW1) detects the number. It collects the digits of the number and sends them to the MGC to make a decision (Figure 5.3, step 2).

- The MGC asks GW1 to create a new connection (Figure 5.3, step 3). The session description information is sent in the ACK to the createConnection message.

- The MGC determines the call routing information using external sources (for example, E.164).

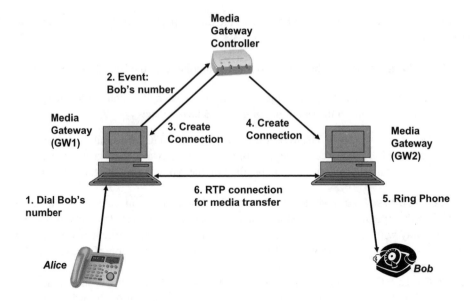

Figure 5.3   A basic MGCP-based call setup operation.

- The MGC orders GW2 to create a connection and sends it the required parameters (Figure 5.3, step 4).

- On MGC's command GW2 rings Bob's phone (Figure 5.3, step 5).

- Based on the session description parameters in the ACKs, MGC asks GW1 and GW2 to create an RTP connection, thus creating the voice path between Alice and Bob.

This is a very simple scenario we have described. In general the message exchange process is more complicated with several nuances addressed in the standard. For more details on MGCP, the reader is referred to IETF RFC 3435 [3].

In a similar direction, ITU and IETF are jointly proposing the Gateway Control Protocol/H.248 (earlier version called Megaco) which takes MGCP to the next level. For more details the reader can refer to the corresponding IETF RFC 3525 [4].

5.3   MEDIA TRANSPORT PROTOCOLS

The second type of protocol involved in transportation of voice are those that actually carry the digitized voice in form of packets. The most widely used protocol for this purpose is the Real-Time Transport Protocol (RTP). We give a brief overview of this protocol.

5.3.1   Real-time Transport Protocol (RTP)

The traditional transport protocols are User Datagram Protocol (UDP) and Transmission Control Protocol (TCP). UDP is a unreliable data transport protocol meant to deliver packets to the destination without worrying about whether or not the packets were actually delivered. A packet dropped in between by some router on the path is not retransmitted by UDP. TCP, in contrast, is meant for reliable delivery and will retransmit a lost packet so as to compensate for the loss. These protocols do not worry about the latency of the data delivered.

RTP is meant to transport data packet with real-time characteristics. RTP can use both TCP and UDP as a transport layer but in most cases it uses UDP. VoIP packets need to be delivered in real-time since delayed packets become meaningless after a short while. Thus, using TCP that will retransmit the lost packets offers little gain. RTP over UDP serves as an excellent vehicle for these packets. In general, RTP is used in transporting both voice and video data. However, RTP on its own does not give a guarantee of timely delivery of data, i.e. the packets may be delayed despite the use of RTP. The fast delivery of packets is the responsibility of the lower layers such as UDP/TCP and possibly of the network services that may provide extra QoS support to VoIP packets.

RTP facilitates the transport of real-time data by allowing time-stamping of packets. The actual media forms the payload of an RTP packet. The header contains control information including the encoding type used in the packet, the sequence number of the packet, and the time-stamp of the packet. Furthermore, the receiver is allowed to receive streams from multiple sources. The receiver can arrange the packets in sequence using the time-stamps and sequence numbers in conjunction with the synchronization source field. Note that the sequence number or time-stamp does not suffice on its own for the receiver to reconstruct the stream. Sequence numbers merely give the order in which the packets were generated but give no idea about how far apart in time they were. Conversely, the time-stamps give the instants on which those packets were created but give no sense of continuity as any packets dropped in between cannot be accounted for.

An interesting feature of RTP is that it allows applications certain control through the use of Application Level Framing (ALF). Each application can define its own specific extensions for RTP using profile specification documents. Of course, all parties involved in an RTP session have to agree upon the extension. There is 1 bit in the RTP packet header indicating that there is a header extension which is application-specific. More details about RTP can be found in the corresponding IETF RFC 3550 [5].

5.4 SUMMARY

Successful operation of VoIP depends on communication protocols that help set up the end-to-end communication path between the endpoints and transport the voice data between them in real-time. There are several signaling protocols to set up the call correctly. Of these, SIP and the H.323 set of protocols are very rich and provide end-to-end signaling. SIP uses SDP as its underlying capability exchange mechanism. For system control, H.323 uses RAS to allow terminals to communicate with the gatekeeper, Q.931 to set up the call and H.245 for capability negotiation. MGCP acts as an internal protocol, making a distributed gateway implementation appear as a single gateway to the outside world. RTP is the de-facto standard for media transport.

REFERENCES

1. Rosenberg, J., Schulzrinne, H., Camarillo, G., Johnston, A., Sparks, R., Handley, M. and Schooler, E. SIP: Session Initiation Protocol, *IETF Request for Comments RFC 3261* (2002).

2. Handley, M. and Jacobson, V. SDP: Session description protocol, *IETF Request for Comments RFC 2327* (1998).

3. Andreasen, F. and Foster, B. Media Gateway Control Protocol (MGCP) Version 1.0, *IETF Request for Comments RFC 3435* (2003).

4. Groves, C., Pantaleo, M., Ericsson, L.M., Anderson, T. and Taylor, T. Gateway control protocol Version 1.0, *IETF Request for Comments RFC 3525* (2003).

5. Schulzrinne, H., Casner, S., Frederick, R. and Jacobson, V. RTP: A transport protocol for real-time applications, *IETF Request for Comments RFC 3550* (2003).

PART II

VOIP IN OVERLAY NETWORKS

The goal of this section is to describe the emerging area of VoIP in Overlay Networks. The emergence of P2P telephony applications such as Skype has been possible due to the technological advances in the field of overlay networks and its subfield of P2P networks. This chapter aims to demystify the working of such networks and subsequently the issues arising in using VoIP applications in this domain.

In this section, we will look first at the technology behind overlay networks for an insight into what the Internet can and cannot do and the void that overlay networks fill. After explaining the features of overlay network in Chapter 6, we detail a subset of overlay networks called the P2P networks in Chapter 7. These two chapters establish the fundamentals of the overlay network and the P2P network technologies. We proceed to delineate the operation of VoIP in the general overlay networks in Chapter 8 where we focus on the issues in the data plane routing of VoIP calls in the overlay network. The control plane functionality of VoIP calls is the key point to address in the P2P overlay networks. We elaborate on this aspect in Chapter 9.

At the end of this section, the reader will be well-versed in the fundamentals of overlay networks and the issues arising when they are used to support VoIP.

VoIP: Wireless, P2P and New Enterprise Voice over IP   Samrat Ganguly and Sudeept Bhatnagar
© 2008 John Wiley & Sons, Ltd

CHAPTER 6

OVERLAY NETWORKS

P2P technology has facilitated the deployment of several new applications in recent times. These advances have relied on the advances in the field of overlay networks. The goal of this chapter is to elaborate on the concept of overlay networks and highlight the developments in this field. The insights obtained from several research and commercial efforts in the field of overlay networks have been successfully utilized in the P2P arena. We believe it is important to understand the underlying technologies that have led to the proliferation of P2P applications before delving specifically into the P2P application area and the specific area of VoIP in these networks.

6.1 INTERNET COMMUNICATION OVERVIEW

It is important to understand the working of the Internet to gain an insight into the emergence of overlay networks. The proliferation of overlay networks is largely to address the shortcomings of the current Internet. The fundamental architecture of the Internet carries these limitations because during its infancy, the applications it was designed for were very simple and the underlying hardware had limited capability. Many of the applications today could benefit from a more flexible architecture. However, due either to the architectural limitations underlying the Internet or to the current business model, these desired features are not available to the applications. We give a brief overview of the Internet operation and use this description to reason about the need for overlay networks.

VoIP: Wireless, P2P and New Enterprise Voice over IP   Samrat Ganguly and Sudeept Bhatnagar
© 2008 John Wiley & Sons, Ltd

6.1.1 Communication operations

At a high level, any communication over the Internet can be thought of as being composed of two processes:

- *Resource Discovery:* The process of resource discovery involves an end-user locating the resource with which it wants to communicate. For example, when we visit a web page in our browser, we use DNS to determine the actual IP address of the machine where that URL is hosted. We can expand the connotation of resource discovery to include content discovery where we use a search engine to find the web pages that have some content that matches our interest.

- *Content Transfer:* The content transfer phase involves movement of data between the discovered resource and the end-user. The process of content transfer is not just unidirectional but also includes two-way communication where the end-user can interact with the remote resource interactively. For example, a simple FTP session involves a unidirectional transfer of a file from one machine to another. On the other hand, a VoIP session between two machines involves data being transferred in both directions.

Notwithstanding the other required features (such as the ability to handle hosts behind NAT/Firewalls both for discovery and content transfer), these two operations form the core of Internet communication. Thus, any communication architecture must support these two operations to supplement a wide range of applications.

6.1.2 Communication roles

In the Internet, the communication architecture provides these functionalities by assigning distinct roles to different nodes forming the network. Each role specifies the actions that that particular role is expected to perform in the Internet. Collectively, the entities enacting these roles perform the resource discovery and the content transfer functionalities. There are three primary roles in the Internet:

- *Client:* A client is an entity that originates a communication in order to request some service. Conventionally, all end-hosts in the Internet are clients.

- *Server:* The server is the entity that responds to a client's request by transferring to it the appropriate content.

- *Router:* A client and a server may not be directly connected to each other. A router's role is to forward a packet towards its destination.

Fundamentally, client and server are the primary roles in the Internet that form the endpoints of communication. The routers serve the purpose of relaying packets between them. Along with these three roles, there are other auxiliary roles that help the client and server communicating. For example, a cache stores the objects and acts as a server to the clients in order to reduce the load on the server and reduce the perceived response latency at the client.

6.1.3   Internet routing

The Internet is built on top of the IP layer. When an end-host needs to communicate with another node connected to the Internet, it sends data in the form of a packet and stamps the IP address of the destination in the packet's header. The IP routers in the Internet look up the destination IP address in their local forwarding tables, and accordingly forward the packet towards the next node leading to the destination. The Internet routing mechanism ensures that by following the forwarding tables in the routers in the network, the packet will reach its intended destination. One of the key reasons for the scalability of the Internet is the design of the routing mechanism where each router does not need to know the entire route to each destination.

The routing structure in the Internet is hierarchical. Along with improving the scalability of the network, the hierarchy caters to the business requirement of having separate interoperable network domains under separate owners (the ISPs). Thus, routing across a single domain follows an intra-domain routing protocol such as OSPF or RIP and the routes over multiple domains are constructed using an interdomain routing protocol such as BGP. The Internet only promises a best-effort service to the user without giving any guarantees regarding the QoS that the packet will receive. Thus, the end-users do not have any control over how their packets are routed or processed in the network.

6.1.4   Client–server architecture

Traditional Internet-working is based on the client–server architecture. In this architecture, the client requiring a particular service contacts a predefined server node in the network. The server processes the client's request and takes appropriate action. An example of this type of interaction is the World Wide Web where an HTTP server responds to a client browser's request to GET a particular Uniform Resource Locator (URL) by sending it the content of the corresponding web page.

In the client–server architecture, there is a clear separation between the service provider (server) and the service consumer (client). The service provider maintains the infrastructure comprising several machines that act as servers. In reality, the term server has become synonymous with the machine that performs the server role for any communication in the client–server model. Similarly, the infrastructure for supporting services such as DNS and caching is distinct from the client. Thus, the end-hosts always play the role of the client, the servers are hosted by the service providers (e.g. content providers), and the ISPs perform most of the routing. This separation of roles is amenable to the business model where the ISP can charge for the Internet access service, the content provider charges for access to its content, and the client pays for accessing the content over the Internet. However, this limits the client's ability to improve its performance using its application-specific knowledge.

6.2   LIMITATIONS OF THE INTERNET

The Internet in its current form has become heavily ossified. The simple best-effort service model exported to the end-user severely undermines the application's ability to use its own knowledge of the network requirements to tune the network path to attain the desired effect. For example, the network essentially treats the packets of an interactive telnet session and a bulk-data transfer using FTP identically. The Internet architecture is fundamental to some of these limitations and the business model causes the others. We will discuss some of these limitations using the Internet architecture detailed above.

- *Routing:* As discussed above, the hierarchical routing in the Internet helps in attaining scalability and isolation of different network domains under the control of different ISPs, as is the path between a client and server, furthermore. However, this hides the path characteristics from the client, and the client is bound to use the path determined for it by the Internet routing. For example, if a client were interested in a high bandwidth path rather than a minimum hop path to the server, it does not have a way of choosing a path.

- *Multicast:* Several group communication applications such as audio/video-conferencing require that same data to be sent to multiple endpoints. This can be efficiently achieved by using the well-studied IP multicast mechanisms. However, these mechanisms have not been enabled by the ISPs. Thus, group communication applications are reduced to doing multiple multicasts over the Internet.

- *QoS:* Real-time applications require that their packets be delivered within some acceptable latency. QoS research has established several mechanisms to provide desired guarantees to applications. However, due to lack of ISP interest, this capability is not available to applications.

- *Anycast:* The Internet architecture does not provide any mechanism to the service provider to direct a particular client's request to an appropriate server from among a pool of servers. For example, a service provider would like to send a client's request to the server physically closest to it. Such a choice is not available inherently in the Internet architecture.

- *Privacy:* The single-path routing mechanism of the Internet creates privacy issues. A packet originates from the source node and has the destination IP address stamped in it. Each router along the path knows that the source node is contacting that destination. The close coupling of endpoints to the flow results in this limitation.

While the aforementioned list of limitations is by no means complete, it shows that Internet architecture leaves much to be desired. It has become increasingly difficult to deploy and test new services and applications that go beyond the traditional network service model. The network service providers are reluctant to deploy new functionalities in the routers that may enable new applications. Several technologies have been researched heavily but have failed to take off in the real world for one reason or the other. Thus, testing new network-service ideas on a large-scale real-world network requires a new type of network. It is this void that motivated the emergence of overlay networks.

6.3   OVERLAY NETWORKS

Overlay networks have emerged as an answer to the shortcomings of the Internet. Several limitations of the Internet can be addressed using an overlay network. Before delving into the details of how applications have used overlay networks to bypass these limitations, we first define what is an overlay network.

Definition: *A network built over another network is called an overlay network.*

This definition is very broad. In fact, it includes the earlier version of the Internet itself which was an overlay on top of the telephone network since the nodes communicated over the telephone links using modems. For our purpose, we are interested in the network of nodes that are connected to the Internet but act cooperatively as a network themselves. Figure 6.1 shows an instance of an overlay network where the end-hosts E_i combine to form an overlay network over the network that additionally consists of the routers R_j.

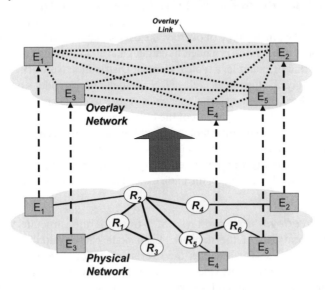

Figure 6.1   Overlay network.

While the nodes can be far off in terms of physical distance and the Internet routing path between them can span multiple hops, in the overlay network they are deemed to be connected by one overlay link. This overlay link is a concise representation of the underlying Internet path.

Definition: *A logical link that denotes the multihop physical path between two overlay nodes is called an overlay link.*

The overlay links between each pair of nodes never cease to exist. The status of the link keeps on changing as and when the underlying physical links change. In the extreme case, a physical link on their overlay link can fail and they will be connected with a new path (with potentially different characteristics) after the Internet routing converges.

Monitoring the overlay links to track their characteristics is much more challenging than a physical link. This can be attributed to the fact that its characteristic depends on the dynamic characteristic of several physical links. Furthermore, several overlay links can share physical links. This causes their characteristics to be interdependent. For example, in Figure 6.1, overlay links $E_3 \rightarrow E_1$ and $E_3 \rightarrow E_2$ share the physical link $R_1 \rightarrow R_2$. In fact, it is possible to have shared links between overlay links that have no common endpoint.

6.3.1 Types of overlay network

In general, overlay networks provide a new way of deploying new services. Consisting of a bunch of end-nodes that act as a network on top of another physical network, an overlay network can enable several new services. The role(s) of each of the overlay nodes depends on the type of communication architecture that the overlay network is supposed to provide. Depending on the types of roles that the overlay nodes play, we can classify the overlay networks into two categories: (a) Infrastructure Overlays; (b) P2P Overlays. The salient issues that arise in the two variants are discussed next.

6.3.1.1 Infrastructure overlays

In an *Infrastructure overlay*, the overlay nodes primarily perform the roles of server and router. The clients are some nodes that are *external* to the overlay. In this type of overlay, the overlay nodes can be thought of as being under the control of one service provider, having high availability and connected to a high bandwidth network. For example, a typical Content Distribution Network (CDN) forms an infrastructure overlay. All the nodes in the CDN are under the provider's control and act as servers/caches. The clients are external machines whose requests are served by the CDN. The clients themselves do not form part of the overlays.

Infrastructure overlays still maintain the clean separation between the service provider and consumer. The added flexibility available to the service provider is that the client requests can be routed to appropriate servers using some knowledge that is available solely to the service provider. For example, the choice of server can be based on the proximity to clients, load on servers, etc.

6.3.1.2 P2P overlays

P2P overlays differ from the infrastructure variants in that the clients also form part of the network. Each node can essentially perform any of the communication roles. This enables the overlay to become infrastructure-free from the perspective of the service provider since now the end-hosts themselves perform the roles of the servers. While the nodes perform the roles of routers as well, the need for ISPs still remains since the overlay routers operate on the overlay links which depend on multi-hop physical paths (provided by the ISP). The popularity of the P2P paradigm emanates primarily from its infrastructure-free service deployment characteristics. Thus, any application meant for a large user-base can leverage the corresponding P2P overlay network without having to deal with setting up a large server base. This opens up a great opportunity to serve as testbed for novel ideas which could not have been implemented in the Internet without a large investment.

6.3.1.3 Design considerations for infrastructure versus P2P overlays

Since the two flavors of overlay primarily differ in the client participation, they may not seem to have a major impact on the involved operations. However, the characteristics of the clients create unique problems that need to be addressed when designing algorithms.

The construction of infrastructure overlays can be optimized using the knowledge that all the overlay nodes are trustworthy and the concern of maintaining client privacy is not there since each client interaction is isolated from others. The design can also leverage the fact that all the nodes are highly available as they are meant to provide some service. On the other hand, a service designed to operate on top of a P2P network needs to handle the high churn that is likely when end-users are also part of the overlay. This is because the end-hosts can often be turned on/off by their users. Lastly, an infrastructure overlay is likely to have high bandwidth network connection. Therefore, the design does not have to deal with the heterogeneity of resources as is typical with end-hosts where users can connect using fiber-to-the-home as well as a slow modem. These differences in availability, capability, and security requirements lead to different solutions for system design over these platforms.

6.3.2   Routing in overlay networks

In any network, there is a need for a routing mechanism that enables two endpoints to communicate over multiple hops. This is attained in the Internet using routing protocols which populate the forwarding table at the routers and the routers forwarding packets using purely destination IP address stamped in the header. Routing in overlay network is performed over the overlay links between the overlay nodes. The network has no control over how the packets are routed in the physical network that connects two overlay nodes. The routing determines the sequence of overlay nodes that a packet should follow before reaching its destination.

In an overlay network, the routing is more challenging than in the Internet. The foremost of these challenges comes from the topology on which these networks operate. In a real network, the nodes (routers and hosts) are connected to other nodes via physical links. Thus, their neighborhood consists of very few nodes. The routing process at a node merely decides to which of its neighbors the packet should be forwarded. However, in an overlay network, the nodes are connected by overlay links that are logical paths constituting a series of physical links. Thus, in this case, each overlay node can potentially have a link to any other overlay node, making it a fully connected structure. The problem in this case is not just the computation of the routes but also to identify the topology on which it should operate. If we treat the fully connected graph as the topology, then the routing between any pair of nodes is essentially the direct overlay link between them. However, the routing updates that are required to keep nodes updated regarding the status of their neighbors would make routing on a fully connected network unscalable. Therefore, a healthy balance is needed between the efficiency and the overhead to attain it.

The second major challenge pertaining to overlay routing comes from the availability of several possible routing metrics. In the Internet, the application has no control over the route its packets take to a destination. The routing is implemented using the policies of the service providers rather than catering to the application need. However, depending on the application that the overlay supports, the routing mechanism can create different routes. For example, an overlay used for streaming media can use low latency paths where as an overlay used for file transfers it can use one or more of the high bandwidth paths. This flexibility in the choice of path is heavily explored in the research community.

Another aspect of overlay routing that needs to be mentioned is that, depending on the characteristics of nodes involved in the overlay network, the design choice can vary significantly. For example, if the overlay nodes join/leave the network frequently, the routing algorithm needs to adapt quickly to reroute any paths that were using those nodes. While use of such *restoration paths* is possible in a subset on Internet domains (those that use MPLS/ATM-based virtual circuits), it is not possible in the traditional networks where there is a single minimum cost path. Furthermore, even if the creation of restoration paths is possible, the process of path failure (mandating the use of alternate paths) is infrequent as it only happens when equipment fails. In instances where the overlay nodes are run-of-the-mill home computers, the users can turn the computers on/off frequently, making the routing change a norm rather than an exception.

As we will see later, the choice of metric for routes varies based on the application. In fact, in the case of a P2P overlay network, the resource discovery phase is tightly coupled with the routing. Thus, applications govern routing in the overlay network, unlike in the Internet where the routing is an application-independent service.

6.4   APPLICATIONS OF OVERLAY NETWORKS

Overlay networks enable rapid deployment of new network applications and services. Fundamentally, an overlay network can behave like any other network. An application that needs to be deployed on a network can use an overlay network as the first step to full-scale deployment. These applications exploit the availability of diverse roles on the network nodes to create new services. We list a few applications that use overlay networks to their benefit.

6.4.1   Content distribution network

Perhaps the most well-known instance of an overlay network is a Content Distribution Network (CDN) such as that of Akamai. The purpose of a CDN is to off-load the task of handling the end-user's content request from the content provider. To this end, a CDN consists of a large number of nodes (overlay CDN nodes) that act cooperatively as an overlay network. The location of these nodes and the size of the network are based on the expected user population that the network is to serve.

A content provider generates content and delegates the responsibility of delivery to the CDN. When a user requests some content from the concerned provider, the request is redirected to one of the CDN nodes using some transparent mechanism. The CDN uses its central algorithm (potentially for load-balancing and response latency minimization) to determine which of its nodes should answer the request and redirects the request (or subparts of the requests) to these nodes. Subsequently, those nodes fulfill their assigned request and the end-user receives the desired content through the CDN. For example, consider an organization X has a contract with a CDN Y to deliver its web-content. When a user visits http://www.X.com in the browser, the browser would generate a DNS request to this URL to find the web-server's IP address. The DNS server that acts as authority for www.X.com can return the IP address of a server from Y in its response; so the browser would contact the server from Y for the actual content for www.X.com. This simple example illustrates the use of DNS redirection for off-loading the content delivery task to a CDN. Note that this is the overlay-based implementation of the anycast service that is lacking in the Internet.

Internet Shortcoming: The Internet does not have an inherent ability to allow the service provider to optimize its resource usage. For example, availability of the IP anycast service at the IP level could have provided the desired mechanism.

Insight: Overlay Network can be used to provide the anycast service. This can help in reducing perceived client latency and load-sharing among servers.

6.4.2   Overlay multicast

Overlay multicast technology enables multicasting of content to multiple end-hosts by building a distribution tree on an overlay network of end-hosts. While IP multicast would eliminate the scenario where duplicate packets traverse a physical link, overlay multicast eliminates the flow of duplicate packets over an overlay link. This is implemented by creating a multicast distribution tree on top of the overlay network and requiring the overlay nodes on the interior of the tree to replicate all incoming packets to all its children in the tree. Of course, multiple overlay links can use the same underlying physical link, thus requiring multiple transmissions of the same packet on that physical link. However, its performance is significantly better compared to distributing the entire content using multiple unicast where the source sends the content to each node separately. Not only are the number of duplicates over the physical network reduced, the burden of transmitting the data to the group of users is shared among the users rather than putting the entire load on the source. For example, consider the application-level multicast tree for the scenario shown in Figure 6.2. Overlay node E_3 is the root of the multicast tree. It sends the packets only to E_1 and E_4. The packet-forwarding load is shared by E_4 which replicates and forwards packets to E_2 and E_5. Note that in this scenario, packets do travel the same physical link multiple times in some cases (e.g. $R_1 \rightarrow R_2$ is traversed twice when E_3 sends packets to E_1 and E_4.) However, it is much better than E_3 sending the packets to all destination nodes in which case it would have to have four copies of packets over the $R_1 \rightarrow R_2$ link.

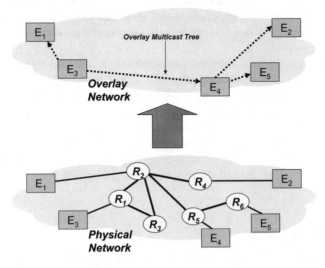

Figure 6.2   Multicasting in an overlay network.

Internet Shortcoming: Group communication applications can benefit significantly by using native IP multicast technology. The Internet business model does not provide an incentive to the ISPs to enable IP multicast.

Insight: A group communication application can scale when users share their processing and bandwidth resources. The multicasting capability is enabled by utilizing the overlay nodes not just as endpoints but also as packet-replicating routers.

6.4.3   Anonymous data delivery

A collection of overlay nodes can cooperate to act as an anonymous data delivery network. The need for anonymous data delivery primarily arises from a desire to enable privacy and free speech on the Internet. When a user normally requests an object from a server, her IP address can be used to associate the individual with the object. Furthermore, traffic analysis can correlate the user with frequented Internet sites revealing sensitive information. In anonymous data delivery, the objective is to deliver the information to the user (and likewise publish information from a user) while protecting the user's identity.

One use of overlay networks is in building such anonymous data delivery networks. The fundamental idea in such networks is to blur the line between the IP address requesting the content versus the IP address that is merely forwarding the content. A message originating at node X is forwarded over multiple overlay nodes in the network. Each forwarding node merely knows the identity of its previous and next hop in the message. A message sent in response to the message from X would be sent along the reverse path. The key point is that no node on the message path is privy to the entire end-to-end sequence of nodes taken by the message. This mitigates the possibility of distinguishing between a node being the destination of a message or merely a forwarding entity.

Examples of such networks include Freenet and Tor. The goal of Freenet is to provide free speech capability whereas Tor is meant to limit the vulnerability of sensitive information to traffic analysis. Both use random multi-hop paths through the network for request routing. In addition, Freenet also has a storage distribution mechanism to limit the ability to identify the source of a document. It can slice a file into chunks, store the chunks at different nodes in the network and retrieve the file chunks using their hash-IDs. Since a file can be stored anywhere (and potentially at multiple places) in the network, there is no easy way to identify the owner of that content.

Internet Shortcoming: Any Internet connection is characterized by its endpoints. The source and the destination know each other and so too do the intermediate routers (and anyone snooping on the packets). The coupling of endpoint IP address to a flow in the Internet architecture limits the privacy.

Insight: Overlay networks can enable privacy by hiding the identity of the source and the destination of a message by use of randomized routing. The overlay nodes can act not just as endpoints but also as relays for other endpoints. If there is no distinction between a forwarder and an endpoint, the endpoint becomes anonymous.

6.4.4   Robust routing

Another interesting use of overlay networks is to route around failures and bottlenecks. The instability in Internet routes is further exacerbated by the policies and the convergence latency of the routing algorithms. To overcome this limitation, an overlay network can be created where each overlay node acts as a router and monitors the quality of the overlay links to a subset of overlay nodes. The idea is that if a path from node A to B becomes unusable due to some physical link failure or congestion, all packets from A to B can be routed through some overlay node X such that the paths A to X and X to B are in good shape. The choice of the relaying router X depends on the setting. For example, if the purpose is to route around failure but keep the minimum latency, X can be a node so that the sum of latency from A to X and X to B is minimum. On the other hand, if the goal is to route around congestion but to have a high bandwidth path, then X should be the node that has the maximum bottleneck bandwidth along A to X and X to B. In the extreme case where a link on the path from node A to B fails, we could say that routing using overlay can allow A and B to communicate whereas they could not have communicated over the Internet until the routing algorithm stabilized and chose a new path.

Internet Shortcoming: The Internet routing mechanism is slow to react to congestion and failures. Moreover, the use of simple routing metrics, such as hops, along with not allowing the endpoints any control over the path, limits a flow's ability to choose a path that meets its requirement (in terms of latency or throughput).

Insight: Overlay networks enable fast rerouting and recovery to bypass bad paths by relaying traffic. Moreover, routing through overlay nodes, the endpoints can select better paths in terms of latency and throughput.

6.4.5 High bandwidth streaming

Another type of service that an overlay network can deliver is high-bandwidth data transfer. Fundamentally, this involves choosing an overlay node X that has good bandwidth characteristics to both the source and the sink. If the minimum of bandwidth from source to X and from X to sink is higher than that offered by the direct physical path between them, the effective data transfer rate between the nodes will be higher when using the overlay node as a relay.

In addition, the underlying Internet routing ensures that there is exactly one path between the source and the sink. Therefore, if the bottleneck on that path is not the access link to the source or the sink, the full capacities of the source and the sink are not being utilized. If there were multiple paths, each having a different bottleneck, between the source and the sink, the transfer capacity would increase significantly. Ideally, we would want the data transfer rate attained in any transfer to be limited only by the access capacity of either the source or the sink. Overlay networks enable such a high-rate data transfer. By using multiple overlay nodes to relay different portions of the data, the effective transfer bandwidth attained between the source and the sink can be increased.

Along with utilizing multiple paths for attaining high cumulative throughput, streaming media adds further constraints. If a packet is delayed beyond a certain time limit, it becomes meaningless for the application. For example, an audio packet containing data that was supposed to be played back at time t loses its relevance if it arrives at the destination after t. This could happen due to transient congestion on the path that that packet took or in extreme cases due to link failures. In such cases, redundant copies of the packet could be transmitted over different overlay paths and, even if one copy of the packet is received, the playback continues smoothly. This sort of redundant transmission for media streaming is used by content providers such as Akamai. Figure 6.3 shows how overlay node E_1 replicates the incoming stream and sends it over a separate overlay path to reduce the perceived packet loss rate.

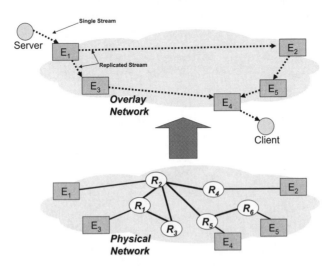

Figure 6.3 Using multipath forwarding in an overlay network.

Internet Shortcoming: Each flow is constrained to use a single path. In the network, there are multiple paths that can be used to connect to an endpoint.

Insight: Spatial diversity of Internet paths can be leveraged to reduce the cumulative transfer latency seen by a user and to conceal packet loss. The use of overlay nodes to relay traffic makes this possible.

6.5 SUMMARY

The Internet architecture has some fundamental shortcomings that limit its capability to support several applications. Overlay networks have emerged to allow applications a good trade-off between flexibility and performance. The routing in overlay networks has to account for the types of node involved in the network and in extreme cases to deal with high user churn and a high degree of resource heterogeneity. Routing in overlay networks is determined by the application it supports.

Several new applications have utilized the flexibility offered by overlay networks to their advantage. These applications have shown that overlay networks provide unique characteristics that can help applications immensely. Some of these features are:

- Overlay networks can be used to provide the anycast service to choose appropriate servers for different clients.

- Overlay networks can enable multicast by using the overlay nodes not just as endpoints but also as packet-replicating routers.

- Overlay networks can enable privacy by hiding the identity of the source and the destination of a message by use of randomized routing.

- Overlay networks enable fast rerouting and recovery to bypass bad paths by relaying traffic.

- The spatial diversity of Internet paths can be leveraged to reduce the cumulative transfer latency seen by a user and to conceal packet loss.

CHAPTER 7

P2P TECHNOLOGY

P2P technology has emerged as a basis for several new applications. Along with VoIP, these applications include general-purpose storage (OceanStore), media streaming (CoolStream) and multicast (Scribe). However, the application that led to the immense popularity of the P2P model is its most controversial application: file-sharing. To understand how this technology has been used in the context of VoIP, it is important to gain an insight into the basic working and the knowledge garnered in this area by different projects.

Continuing from the previous chapter, which gave an overview of the P2P domain, this chapter delves into the emergence of P2P technology over the last few years. The goal is to understand the concepts and features of the P2P technology itself. To this end, we discuss different projects that have used and evolved the concepts of P2P technology, albeit in different contexts. The key focus of this chapter is to understand resource discovery in P2P networks.

We first detail the P2P architecture specifics and contrast it with respect to the conventional communication architecture. Next, we discuss the technological advancements that enable the deployment of these P2P architectures.

7.1 P2P COMMUNICATION OVERVIEW

As discussed earlier, there are two key operations in any network communication: resource discovery and content transfer. While content transfer in most P2P settings is similar to that in the conventional Internet architecture, the resource discovery phase is quite different.

VoIP: Wireless, P2P and New Enterprise Voice over IP   Samrat Ganguly and Sudeept Bhatnagar
© 2008 John Wiley & Sons, Ltd

In addition, a P2P network also has to deal with nodes joining and leaving the network in a manner different from the traditional Internet. The reason for this stems from the fundamental properties of a P2P network, the properties of which we discuss in detail.

7.1.1 Peer node

A node in a P2P overlay network serves all the primary roles: client, server and router. Any node participating in a P2P network is called a *peer*. The node itself is typically an end-host computer connected to the Internet. The node may have few resources in comparison to a server residing in an infrastructure or a router in the Internet. The computing power, storage capacity and network access bandwidth may be limited, and the bandwidth too fluctuating over time. The node may not remain connected to the Internet at all times. For example, the computer user may turn off the computer during the night. Thus, the capabilities and the availability of a node in a P2P network may be limited. In fact, the peer nodes can be very heterogeneous. Some nodes can be high-end having high power CPUs (3–4 Ghz by current standards) whereas others can have CPUs which may be a decade old (200 MHz). Some nodes may have huge storage capacity (several GBs of RAM and several hundred GBs of disk capacity by current standards) and others may have very little (a few MBs of RAM and a few GBs of disk capacity). The access capacity can vary from several Mbps (fiber-to-the-home) to a few Kbps (modems) and the duration of connections could vary from a few minutes to all day.

This diversity of capabilities and availability of peer nodes leads to the challenges that arise in P2P networks. Any solution based on the P2P design needs to account for this diversity. At the same time, the solution needs to perform well so as to encourage peers to use it. This leads to unique challenges and innovative solutions.

7.1.2 Node join and leave

In a P2P network, the process of node joining and leaving also needs special attention and can be different from the conventional techniques. At a high level, this is because a peer node serves all roles in the network. Thus, it simultaneously joins the network in the capacity of client, server and router. Thus, depending on the organization of the network, the node has to be added in all three roles. Similarly, when a node leaves a P2P network, it disrupts the network's functioning in all three roles. Perhaps leaving as a client does not have much impact on the network, but leaving as a server and a router can affect the network service and the connectivity. Thus, joining and leaving of a node needs to be handled as a separate task.

We shall see later in this chapter that the node joining and leaving processes are not generic. Depending on the organization of the P2P network, they can vary.

7.1.3 Bootstrapping

Bootstrapping in a P2P network refers to the process of a node finding the identity of a rendezvous node that is already inside the network. The node-joining process mentioned earlier follows after the new node contacts the rendezvous node. Typically, if a peer leaves a P2P network (say the computer is turned off), it stores the locations (IP addresses) of a few other peer nodes in the network. If it decides to rejoin the network, it tries to contact the peer nodes whose addresses it has cached. If any one of them is still in the network, the node can rejoin. However, in case none of those nodes is in the network anymore, there is a need to provide an out-of-band mechanism to allow the node to join the network. The same happens when a node joins the network for the first time.

The bootstrapping mechanism to join a P2P network typically involves the use of well-known servers that a user contacts as a last resort. These servers keep an updated list of peers who are in the network. For example, in the Gnutella network, the GWebCache caching systems store the active peers in the network. Any node can contact an existing cache to find out a list of peers for bootstrapping.

7.1.4 Communication process

Similarly to the conventional client–server architecture, in a P2P network as well, the client initiates the communication process. While in the client–server architecture, the server is usually known and the routers typically have stable paths from the clients to the servers, this is not so in a P2P network. The fluctuating availability of the peer nodes imposes a new problem. The nodes that act in a server role can keep changing with time. A client's request to discover a particular service can lead to different servers at different times. In this sense, the resource discovery phase in P2P is heavily coupled with routing in a P2P network. At a high level, one can view resource discovery as akin to routing in such a network. The goal is to ensure that a client's request reaches the appropriate server. The client does not know the identity of the server and the network helps it find the correct server. In this regard, the process is of resource discovery. On the other hand, the P2P network typically does not have a fixed infrastructure. Thus, the discovery request itself travels over possibly multiple overlay hops to arrive at the node that can answer the client's resource discovery request. This implies routing of the message to the correct destination. Furthermore, the availability fluctuation of nodes also means that the routers keep on changing. Thus, the network itself keeps on evolving over time and yet it needs to ensure that the clients keep finding servers for their requests.

The P2P networks vary in the manner in which they ensure resource discovery (and hence routing). We shall study these methods in detail later in this chapter. Once the client discovers a server (or a set of servers) for the service it seeks, it can start the content transfer from it (them) using an appropriate data transport protocol. For example, if the client is downloading a file from the server, it can use FTP or HTTP and if it is streaming media or having a VoIP conversation, it can use RTP. This process remains similar to the content transfer for the conventional client–server architecture.

Note that routing of the resource discovery request is not the only form of routing in P2P networks. In certain applications, the content being transferred between the client and the server (post-discovery) can also be routed over multiple overlay hops to attain some application-specific objective (such as higher throughput, lower latency, anonymity). However, the route computation in these instances is similar to that over the Internet except that the links are overlay links.

7.2 CLASSIFICATION OF P2P NETWORKS

While we have identified that different implementations of P2P networks may differ in the way nodes join and leave these networks and the way they resource discover, we need a framework to classify these P2P networks to group their mechanisms together. In this regard, we classify these P2P overlays based on the structure that they possess. Note that in a P2P network, the term structure refers to the overlay connectivity and the positions of the nodes with respect to their utility in terms of routing and resource discovery. We classify the P2P overlays into the following categories:

- *Unstructured Overlays:* In these networks, there is no structure imposed onto the network. The peer nodes essentially form overlay links with an arbitrary set of peers. Perhaps the neighboring set of peers could be based on certain heuristics such as connecting to nodes that are proximate (low latency); however, from the perspective of routing and resource discovery, all peer links are essentially equal. There are few (if any constraints) on the connectivity and service placement.

- *Structured Overlays:* This type of overlay represent the other extreme in comparison with unstructured overlays, although in both of them all nodes are treated equally. In structured overlays, a definite topological structure is imposed on the constituent nodes. This is done to facilitate routing and request discovery. When nodes join the network, there is a rigid policy that defines the procedure so that they seamlessly integrate into the resource discovery and routing framework while sharing the load and keeping the system scalable.

- *Semi-structured Overlays:* In these overlays, a subset of nodes has some structure imposed on them while the remaining nodes do not. For example, certain high-end nodes can be prioritized to handle certain important data and can have a higher neighborhood degree compared to other low-end nodes. One way of looking at these overlays is that they classify the nodes into capability-based subsets so that higher capability nodes can take up the bulk of resource discovery and routing responsibilities and the lower capability nodes merely connect to the sub-network created by the high capability nodes. The nodes chosen to be part of the resource discovery and routing framework may be organized in the form of a structured overlay.

Having identified the three classes of P2P overlay, we need to understand the process of resource discovery in these networks. The process of node joining and leaving also needs to be addressed in case of semi-structured and structured overlays. We shall study these next.

7.3 UNSTRUCTURED OVERLAYS

We first study the possible type of resource discovery in unstructured overlays. Note that node joining and leaving in unstructured overlays is not very critical to routing and resource discovery. When a node joins a P2P network, it essentially contacts one pre-determined peer that is already in the network. The identity of this peer may be hard-coded in the node's software. This boot-strapping node may help in authenticating the new node and can inform it of the other peers to which it can connect in the network. This is independent of the resource discovery framework. Similarly a node wishing to leave an unstructured overlay need not be handled in any special way.

As there is no special way in which the resources are distributed in an unstructured overlay, the choice of method that can be used for discovering resources is very rudimentary. There are two main methods that can be used in such a setting, discussed next.

7.3.1 Centralized resource discovery

The simplest form of resource discovery in an unstructured overlay can be attained using a centralized server that stores the directory of all resources. This is obviously a non-P2P solution but is nonetheless important since the first P2P solutions utilized it. The initial idea was to have a hybrid solution that keeps the content transfer infrastructure free but to have some infrastructure for resource discovery.

In this model, each peer node registers with the centralized directory node indicating the resources that it has. A client looking for a particular resource goes to the centralized server to find out which peer nodes can provide it with the desired resource. The client can then contact those peer nodes to access that resource.

The most popular example of this type of P2P network was Napster. The goal of Napster was to help people share music files. The music files did not reside on the Napster servers. Instead, the files remained stored on the computers of the end-users themselves. A user (peer) would share his music files with other peers by listing the files in the directory hosted at Napster servers. A user looking for a particular music file would then search the Napster directory to find out which of the other peers had this file and then would directly download this music file.

This simple file-sharing model initiated the P2P revolution in the general Internet user-base. However, Napster was closed down because it was deemed as facilitating illegal music downloads despite not hosting any of the content. In fact, this event lead to a slew of effort to decentralize resource discovery in P2P networks. While the purpose may have been file-sharing, these efforts represent tremendous technological advancements in the general arena of resource discovery in P2P networks.

7.3.2 Controlled flooding

An alternative to having a centralized directory of resources is to ask all other nodes in the network whether or not they have the desired resource. In a fully connected overlay network where each node is connected to all the other nodes in the network, this strategy would give the desired results. However, in a large P2P network, it is unlikely that each node be connected to all other nodes. Typically, a node is connected to a small subset of the nodes. In such a setting, the resource discovery request from a source node is sent to its neighbors, who in turn forward it to their neighbors and so on. If a node has the desired request, it replies to the requestor with its address. The source node can then access that resource directly.

This main drawback of this approach is that each request essentially goes to all the nodes in the network, hence generating a tremendous amount of overhead traffic. This is controlled using a *time-to-live* parameter in the request. Each node decrements the time-to-live field before forwarding the request to its neighbors. If a node finds the time-to-live to be 0, it discards the request. Thus, the propagation of request is limited to the number of hops specified by the time-to-live. This effectively attains a trade-off between an exhaustive search of the network for the desired resource and the overhead required to fulfill the request.

This type of resource discovery was used in the initial version of the Gnutella file-sharing application. In the initial version of Gnutella, the overhead was even higher, as a node that contained the desired resource did not respond directly to the requestor. Instead, it responded to its neighbor from whom it received the request that in turn did the same and so on until the response reached the requestor. In fact, even with controlled flooding, the scalability of Gnutella was limited. While it was very robust not having to depend on any single node for functioning, the overhead of running this protocol was high. The bandwidth overheads of searching in Gnutella grew exponentially with the number of users. In many cases, this would saturate the access links of slower nodes, reducing their utility. Gnutella subsequently moved onto a tiered search system.

7.4 STRUCTURED OVERLAYS – DISTRIBUTED HASH TABLES (DHTs)

DHTs form one of the key technologies to have enabled the success of P2P networks. As mentioned earlier, any network has to support resource discovery and content transfer operations. DHTs enable decentralized resource discovery. In essence, DHTs are what make the P2P networks truly peer-to-peer. We elaborate the concept, the underlying algorithms, and evolution of DHTs. The goal is to familiarize the reader with the task of searching for information in a distributed manner.

In a P2P network, a great deal of information, both about the peer nodes and the content resident in the network, needs to be stored. In a large network, there is much information to be stored. Thus, not all the information can be stored at a single node. This limitation is made more stringent by the fact that most peer nodes in a large network are typical low-end home computers connected to the Internet using a modest network connection. Hence, we need a mechanism efficiently to partition all the information among the constituent peer nodes. In order to find some desired information in the network, a user needs to locate a peer who has that information. The search mechanism should ensure that this information-searching process is fast, also. DHTs provide the mechanism efficiently to partition the information in the network and to execute the search and retrieval operations. The concept of DHTs has its roots in the well-known information storage and retrieval method of hashing.

7.4.1 Hashing

Suppose we have a set K of arbitrary keys (bit-strings) along with their associated values. We need to store these keys in an array with N slots, so that they can be quickly located when needed. N is referred to as the *range* of the hashing function. Hashing is a method that allows us to perform this storage and retrieval quickly. We use a hash function $\mathcal{H}$ to map each key $k \in K$ to one of the N array slots (identified by their index). We apply $\mathcal{H}$ on each key $k \in K$ to obtain a number between 1 and N and store the key in the corresponding array slot. Whenever we need to retrieve a value corresponding to a key k', we apply $\mathcal{H}$ on k' to obtain the slot number j. If there is a stored value corresponding to key k', then it must be available in slot j; otherwise there is definitely no stored value corresponding to k'. Thus, hashing identifies the location of a particular key very quickly.

There is no guarantee that two keys will never map to the same slot in the array. Thus, the storage mechanism for each array slot must have some data structure to store multiple keys that may map to that slot. For example, each slot could be a balanced binary tree which stores all the keys based on their actual values. The storage and retrieval in this structure would take $O(\log m)$ time if m entries map to that slot. Thus, the total search time would be the time to hash (which is $O(1)$) followed by the time to locate the key in the slot data structure.

The efficiency of the hashing function used is critical to the operation of a hashing table. Suppose a hashing function is heavily skewed so that all keys are mapped into the same slot. In that case, all the searches end up at that same slot and traversing the underlying data structure would be costly. This effectively eliminates any benefits of hashing. Ideally, a hashing function should map at most $\lceil |K|/N \rceil$ keys in each slot. This would provide the maximum possible benefit in key retrieval.

Table 7.1 shows a simple hash function populating a hash table with $N = 10$ slots. The hash function used is $\mathcal{H}(x) = x^3 \bmod N$. We have five key values, 3, 4, 5, 6, 7, which we have to store into the hash table. The key 3 is stored in the slot given by $\mathcal{H}(3) = 3^3 \bmod 10 = 7$. Similarly, key 4 is stored in slot 4, key 5 in slot 5, key 6 in slot 6 and key 7 in slot 3. In this case, we have each key mapping to a unique slot. However, if we have another key 13 to add to the hash table, it would end up in the same slot as key 3, causing a collision.

Table 7.1 Hash values for some keys using two hash functions.

Key (x)	$\mathcal{H}(x) = x^3 \bmod 10$	$\mathcal{H}(x) = x^3 \bmod 9$
3	7	0
4	4	1
5	5	8
6	6	0
7	3	1
13	3	1

7.4.1.1 *Usage in DHT* Applicability of hashing to DHT is straightforward. If we consider each node in the network to represent one array slot in line with the previous discussion on hashing, the hash function essentially maps each value to one node. Thus, identifying the node which must store a particular key boils down to hashing a key value. However, unlike the ability to reach that slot directly, as is possible in the case of an array, the corresponding peer nodes are distributed all over the Internet. In order to store/retrieve a particular key, we need the ability to route the storage/retrieval request to that particular node. The node would reply with the corresponding value or a null value in case that key is not associated with that value.

Thus, essentially adding the ability to route messages between any pair of peer nodes provides the desired distributed information search and retrieval capability while load balancing is attained using hashing. DHTs provide this routing ability.

7.4.1.2 *Limitations with respect to DHT* From the onset, we have mentioned that nodes joining and leaving are the norm in a typical P2P network. This process has drastic implications on the use of hashing in DHT as mentioned above. The use of nodes to identify array slots essentially means that the number of array slots keeps on changing with each peer arrival and departure. Since the purpose of hashing is uniformly to distribute each key value over the available slots, the distribution of key values can change drastically in this case.

Again consider our toy example in Table 7.1. Suppose one of the nodes in the system left, leaving us with $N = 9$ slots. In this case, the hash function used is $\mathcal{H}(x) = x^3 \bmod N$ becoming $\mathcal{H}(x) = x^3 \bmod 9$. When we hash the same five key values, 3, 4, 5, 6, 7, using this new hash function we obtain the corresponding slots as 0, 1, 8, 0, 1. Note that the slots for each of the keys changed. This implies that we have to move all the keys to the appropriate nodes in the DHT if we use this type of hashing to assign keys to nodes.

In a large network, moving a large fraction (almost all) of the keys with node arrival and departure would incur a huge overhead. In fact, even if this movement were feasible from the perspective of the bandwidth overhead, if the nodes continue to arrive and depart while these values are moving, the system may not even stabilize with the mapping of the node ID to the array slot constantly changing.

This undesirable effect is caused due to the tight coupling of the entire network's structure with the network size (which is affected by the arrival departure of even one node). It is desirable to attain the same effect of load balancing as achieved by hashing, albeit in a setting where the node arrival and departure do not trigger change in the entire network structure. This feature is attained using consistent hashing.

7.4.1.3 *Standard hash functions*
Designing hash functions which attain good load-balancing properties and minimize collisions has long remained a challenging problem. One of the most popular hash functions in use has been MD5. Recently, however, MD5 has been losing its popularity, especially when used as a digital signature of a document. The reason is that methods have been developed that can allow the systematic creation of multiple documents that lead to the same MD5 hash value, thus invalidating its use as a signature of one particular document.

The Secure Hash Algorithm (SHA) family of hash functions is catching on in the cryptography community. For the purpose of DHTs, we do not need cryptographically strong hashes but functions that are unlikely to hash two keys to the same hash and keep the hash distribution uniform. For this purpose most of the designed DHTs have used SHA-1 which maps any string to a 160 bit hash value as the hash function of choice.

7.4.2 Consistent hashing

Consistent hashing is a hashing method that prevents large-scale change in the mapping of keys to slots in the event of slot addition or deletion. An alternate definition is that a hashing function is consistent if it re-maps only a small number of keys when its range changes. When using a consistent hashing to map keys to slots in a hash table, only a small fraction of keys are re-mapped when a slot is added or deleted. Typically, if we have K keys stored in a hash table of N slots, then addition or deletion of a slot would result in the re-mapping of only K/N keys on an average. This property of a consistent hashing scheme is called smoothness. Recall that with the traditional hashing, in the event of addition or deletion of a slot, a large number of keys (even all keys) may be re-mapped. Along with providing smoothness, a consistent hash function provides balance where each slot is likely to have a similar number of keys.

It is important to note that consistent hashing was initially defined for web caching. Thus, a complete definition of consistent hashing incorporated the properties of the web-caching systems. The main element of interest is the concept of a view. Consider a caching system with N caches with K items (web pages) being stored in these caches. A client need not be aware of all N caches in the system. The subset of caches which a client is aware is termed its *view*. The purpose of hashing is to ensure that each item is present in some cache under each view. Under this model, the definition of consistent hashing function calls for the function to support two additional properties (other than smoothness and balance): even load and small spread. Even load refers to no node having an unreasonably large number of distinct cached items over all views of which it is part. This is to ensure that each node is responsible for responding to requests for approximately a similar number of items. Small spread implies that a cached item be present at only a few nodes despite the presence of an item being ensured in at least one cache in each view. An example of such a consistent hashing function is to map both the slots and the keys uniformly on the unit circle. In such a case the successor slot for a key is the slot which follows the key's location on the circle when traversing the circle clockwise. In a practical implementation, consider the circle to represent the range of values from 0 to $2^b - 1$ with the value 2^b coinciding with the value 0. This forms a circle with 2^b points evenly spread on it. The keys are hashed by a uniform hashing function $\mathcal{H}$ with a range of 2^b onto the circle and the same function is used to hash the slots (node-IDs). Therefore, if hashes of two slots s_1 and s_2 are consecutive on the circle when traversed clockwise, then all keys k with $\mathcal{H}(s_1) < \mathcal{H}(k) \leq \mathcal{H}(s_2)$ are assigned to slot s_2. It has been shown that this hashing scheme has all the desired properties. An example of this type of hashing function is shown in Figure 7.1, where there are eight slots assigned random 8-bit IDs (corresponding to slot ID range 0–255), and seven values with the hash function used to store the values being $\mathcal{H}(k) = 3k + 211 \mod 256$. Assume that the slot IDs have been computed by applying the hash function on some globally unique IDs assigned to the nodes corresponding to the slots (say, using their IP addresses).

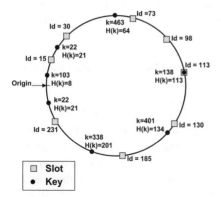

Key (k)	H(k)	Slot-for(H(k))
22	21	30
103	8	15
138	113	113
338	201	231
352	243	15
401	134	185
463	64	73

Figure 7.1   Storing seven key values in eight slots with IDs randomly distributed in the range 0–255.

From the perspective of DHT, the notion of view is not significant and hence the even load and small spread properties are not important. In fact, if all views are identical and cover all the nodes, then these properties can be thought of as converging on the balance property. The properties of interest for a DHT are smoothness and balance. The balancing property of the consistent hashing function clearly helps the nodes in DHTs to participate equally. The DHTs leverage the smoothness property of consistent hashing to reduce the network maintenance overhead in face of churn. Since a node represents a slot in a DHT, when a node joins or leaves a DHT, only a small fraction of the total keys stored in the network are moved around. After the movement of this set of keys, the DHT becomes stable again.

7.4.3   Increasing information availability

Suppose a particular information is stored at exactly one node in the network. If the node leaves the network for any reason, then that information is totally lost. This is equivalent to having a key in exactly one slot and the slot disappearing from the hash table. A traditional solution to this problem is replication. For example, we can use k different hash functions with the same range to hash the keys. Thus, each key is likely to be available in k different slots. In this case, the key will be unavailable only when all the corresponding k nodes left the DHT.

7.5   TYPES OF DHT

There are several DHTs proposed in the recent years. They differ in the amount of routing state required at each of the participating nodes and the corresponding routing hops required for sending a message from one node to another. There are different abstractions they use to define which key resides at which node. We describe a small subset of these DHTs here.

7.5.1   Chord

Chord is one of the most prominent DHTs [1]. Chord essentially uses the consistent hashing function described above where each node and each key is hashed onto a unit circle with b bit identifiers. The circle is called the *Chord ring*. Each key k is stored in the node n which is its *successor* on the ring. For correctness purposes, it is sufficient for each node to keep a pointer (*finger* in Chord terminology) to its successor node on the ring. Thus, search for any key can be completed by nodes continuing to forward a request to their successors until it reaches the node that is responsible for that key. That node can respond to the node originating the query with a value corresponding to the key.

However, while the aforementioned query routing is correct, it is very inefficient as it requires $N/2$ hops on an average to reach the correct node in an N node DHT. To improve the efficiency of routing a message in the DHT, Chord requires each node to keep fingers to $O(\log N)$ other nodes in the DHT in form a *finger table*. The ith entry of the finger table of a node with ID m contains the ID of the *first* node clockwise from it on the ring that has an ID of at least $m + 2^i \bmod N$. Effectively, this node is the owner of the key with the value $m + 2^i \bmod N$. The first entry of the finger table (corresponding to $i = 0$) is also the successor node for m. One may observe that each entry of the finger table points to a node which is at least twice the distance (on the Chord ring) from m than the previous entry in the table. Intuitively, a node keeps more information about its proximate value space than values that are further from it. This is illustrated in Figure 7.2(a) where the finger table for node with ID 15 is shown. The first four fingers correspond to small increments in the large-value space in the Chord ring and all of them end up corresponding to node 30. The longer range fingers are far apart and may skip nodes in between (e.g. finger 7 skipped nodes 113 and 130).

To route a key k in the network, node m uses its finger table. If k lies between the values m and that of the successor of m, then the search is complete and m knows that its successor is the one who is responsible for key k. If, however, k is outside the range between m and its successor, m sends the message to the node (say x) in its finger table whose ID *most immediately precedes* k. This choice takes the message closest to the destination value k as per the information available at node m. Moreover, since a node keeps more information

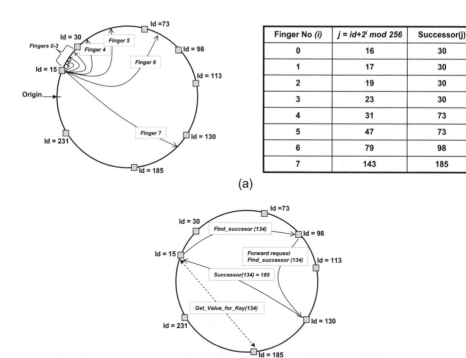

Finger No *(i)*	*j = id+2ⁱ mod 256*	Successor(j)
0	16	30
1	17	30
2	19	30
3	23	30
4	31	73
5	47	73
6	79	98
7	143	185

(a)

(b)

Figure 7.2 (a) Finger table at node 15 in a Chord Ring; (b) hops taken in the search for key 134 initiated by node 15.

regarding the value space closer to its own ID, node x is likely to have more information about the location of the node responsible for value k (successor of k) than node m. This processes continues until the node responsible for value k is reached. Figure 7.2(b) shows the message routing when node 15 initiates a query for key 134. The node first sends the message to node 98, that the largest node (it knows of in the finger table) that precedes the value 134. Similarly, node 98 will forward the request to node 130 which knows that the value 134 lies between itself and its successor node 185. Therefore, it returns the ID (and possibly the IP address) of node 185 to node 15. Node 15 can then contact node 185 directly to transfer the content related to key 134 directly. It has been shown that this type of routing requires $O(\log N)$ hops for an N node Chord ring.

Nodes joining in Chord is simple. When a node m joins the Chord ring, it contacts a random node in the ring m' (which it knows through boot-strapping) to find its successor in the ring. Just by knowing its successor, node m joins the ring. Periodically, each node verifies its predecessor and successor and updates them if a newly joined node is now its predecessor or successor. To keep the correct finger table, each node has periodically to search for the node responsible for the keys that correspond to its finger table. For example, node m would lookup the node responsible for the key $m + 2^i \bmod N$ for each value of i to fill out the correct entries for its finger table.

7.5.2 Koorde

Koorde is an improvement over Chord which is *degree-optimal* [2]. The degree optimality of Koorde refers to the routing in a Koorde DHT requiring the minimum possible number of hops for a given fixed degree d of neighbors that each node can have. Koorde shows how to have a routing structure that requires $O(\log N)$ hops between nodes in an N node DHT while requiring a constant number of neighbors (the base version uses only two neighbors per node). However, the base version of the DHT is susceptible to node failures. To make the system fault-tolerant by keeping a neighbor-list of $O(\log N)$ nodes, Koorde achieves an average path length of $O(\log N / \log \log N)$ which is degree-optimal, also.

Koorde works on a routing structure similar to Chord. The difference between the two is the way the two route a message from one node to another. While Chord stores $O(\log N)$ neighbor pointers at each node to attain an average path length of $O(\log N)$, Koorde attains the same path length using only two neighbors. To achieve this, Koorde uses a well-defined structure called a de Bruijn graph. A de Bruin graph consists of 2^b-nodes with each node corresponding to one number of b-bits. For example, as shown in Figure 7.3, the de Bruijn graph for nodes with 3-bit IDs has eight nodes, one for each label 000 to 111. A node with ID m has edges to two nodes with IDs $2m \bmod 2^b$ and $2m + 1 \bmod 2^b$. One can visualize these two numbers as the two numbers that we obtain by shifting the bits of m to the left by one (throwing away the leading bit) and the last bit being either 0 (for $2m \bmod 2^b$) or 1 (for $2m + 1 \bmod 2^b$). Consider searching a key k (b-bit length) in a de Bruijn network with 2^b nodes so that a node with ID k is responsible for key k. Node m is searching for key k which means that it has to send the message to node k. In order to route the message to node k, node m *shifts in* the value of k one bit at a time. Thus, if the leading bit of k is 0, it sends the key (or message) to the node $2m \bmod 2^b$ and if it is 1, the key is sent to $2m + 1 \bmod 2^b$. The key k is shifted left by one to discard its first bit. The next node does the same with the new left-most bit of k shifted on and so on. After b shift in operations, the entire value of k would be shifted on, implying that the node which performed that last shift-on operation would be sending the key to node k which is responsible for the key. The entire routing takes b hops since b bits are shifted in, one at a time. For example, consider the de Bruijn graph in Figure 7.3. If node 000 wants to search for key 110, it will start shifting in the left-most bit (1 in this case) into its own address, giving a value 001. Therefore, it will send the request to node 001 (to which one of its two edges points). Node 001 will shift in the second left-most bit of the key (which is also 1) in giving the address 011. Again node 001 has an edge to node 011 and will forward the request there. Node 011 will shift in the last bit of the key (which is 0) to obtain 110. Node 011 will send the key to node 110, which is the destination node for this search.

Of course the above procedure is in an ideal case where there is one node for every number. In practice, the number of nodes is likely to be significantly less. To build a DHT out of a fewer number of nodes and yet using de Bruijn graph, Koorde requires that a node keep a pointer to its successor node on the ring, along with the predecessor node to the value $2m$ on the ring. Using these two values, Koorde emulates the traversal from the node m to a key k on an imaginary de Bruijn graph. A key follows the predecessors of all imaginary nodes i that are generated in the b hops taken in the routing on a complete de Bruijn graph. Even in this setting, the expected number of routing hops in the DHT are $O(b)$ while requiring only two neighbor entries. Note that N can be much smaller than b, implying that $O(b)$ is not the same as $O(\log N)$. To attain the $O(\log N)$ number of hops, Koorde uses the idea that node m is the predecessor of all imaginary nodes between itself and its successor.

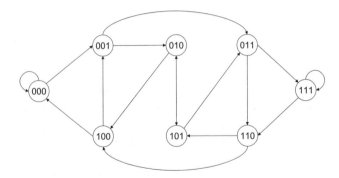

Figure 7.3   The de Bruijn graph with eight nodes.

Thus, node m may shift on more than one bit from k before sending it to the next hop. The number of bits from k that can be shifted on depends on the length of the interval between m and its successor node. It has been shown that with high probability, m can shift on all but $b - 2 \log N$ bits of k before sending it to the next hop. Subsequently, only $O(\log N)$ bits of k remain to shifted on, leading to an expected path length of $O(\log N)$ hops.

7.5.3   CAN

Content Addressable Network (CAN) is a DHT that uses a d-dimensional torus as a basis for assigning keys to nodes [3]. A d-dimensional torus is a d-dimensional hypercube which wraps around its edges. For example, a two-dimensional torus is a donut which is formed by joining the two horizontal edges and the two vertical edges of a square.

The idea used in CAN is that each key can be thought of as a point in a d-dimensional space. Thus, if a node is made responsible for a portion of the d-dimensional space, then all points that lie inside that space will be stored at the corresponding node. Routing a message to a point k in such a case essentially involves routing along the shortest path from a node m to the coordinates of the point. Thus, a node m determines which of its neighbors is the owner of the space that has a coordinate closest to the destination point. Note that the torus shape is essentially of use only for routing and not for assigning co-ordinates. This routing also helps clarify the identities of the neighbors of a particular node in CAN. If a node k has overlapping edges with node m in $d - 1$ dimensions and in the remaining dimension their edges abut each other, then k is a neighbor of m.

When a node m joins a CAN network, it randomly chooses its co-ordinates in the d-dimensional space. It sends a join request to that co-ordinate. The node m' responsible for the corresponding co-ordinate splits the zone it is currently holding into two, assigns one subzone to m, and transfers the corresponding keys resident in that sub-zone to m. The routing table of m is easy to populate as it contains only a subset of the neighbors of m'. Similarly, when a node leaves CAN, its zone is merged with the zone of its neighbor(s).

CAN operates with $O(d)$ neighbors per node and requires an average of $O(dn^{1/d})$ hops to answer requests.

7.5.4 Kademlia

The last DHT that we discuss is Kademlia [4]. The significance of this DHT is that it is currently being used in several popular P2P file transfer applications. These applications include the official BitTorrent client, Azureus (a client implementing BitTorrent), and the Kad Network used by applications such as Emule. Along with ease of implementation, the consistency and performance of Kademlia in fault-prone scenarios are driving its adoption.

While the node IDs and keys are assigned random hash values like the other DHTs, Kademlia differs in its notion of query routing and the routing table structure. The notion of distance between two IDs x, y in Kademlia is the XOR $(x \oplus y)$ of the two values. A node m stores up to k entries in its routing table corresponding to the nodes whose distance is within 2^i, and 2^{i+1} of m. Each such list of nodes is called a $k - bucket$. Unlike Chord where the ith finger entry corresponds to a fixed node, Kademlia's metric implies that all nodes whose ID differs with m in the ith bit with all higher order bits being identical, can form part of the corresponding $k - bucket$.

Routing in Kademlia exploits this redundancy in the number of neighbors to send requests for a key x to k neighbors at the same time. The chosen neighbors are those which are closest to the value x. Since the nodes have more information about the key space closer to their own IDs, these k nodes are likely to have more information regarding the key space in vicinity of x and they return the IDs of the k nodes in their routing table who are the closest to the value x. This procedure converges quickly and is resilient to node failures since a request is sent to multiple neighbors. Also, a Kademlia node can passively use any request/response passing through it to update its $k - buckets$ since each node can be part of some $k - bucket$. So Kademlia can optimize the routing table entries for other metrics such as latency as well.

7.6 SEMI-STRUCTURED OVERLAYS

Resource discovery in the unstructured overlays ended up either being centralized or being too inefficient for the network to remain scalable. This insight made it imperative to impose some sort of structure on the network nodes. The question to be answered is, how rigid a structure can be imposed? Structured overlays use DHTs to impose some structure on *all* the nodes in the network. Thus, each node shared some load in the network. However, for practical reasons it may not be desirable to have all nodes forming part of the routing backbone.

Semi-structured overlays classify the nodes based on their capabilities and their availability. A node with higher capability and higher estimated availability in the network forms part of the routing and resource discovery backbone of the network. These nodes are generally called supernodes or superpeers. Other nodes connect to one such backbone node and route all their resource discovery requests through this rendezvous node. The reason for this discrepancy is obvious – a node that has few resources and is not going to be available for long is unlikely to be of significant help in the resource discovery and routing tasks.

While choosing a node to be part of the network backbone is one part of the solution, the other part is to decide how these nodes participate in the backbone. There has to be a way to partition the resources between these nodes and there has to be a mechanism to route requests between these nodes so that with high probability the request reaches a node which can respond to the request. There exist different methods of attaining these goals and different solutions have taken different approaches in this regard.

7.6.1 FastTrack

Kazaa (and previously Morpheus) uses the FastTrack network for its communication. In the FastTrack network, the clients join by contacting one of the existing supernodes and informing them of the list of files that they are willing to share with other nodes in the network. The supernode then contacts other supernodes in the network and shares this information with them. The specifics of the supernode-to-supernode communication protocol and how they partition the information to be stored are proprietary and not known publicly. The client does not contact other nodes in the network directly for the purpose of sharing the directory of its available files. All client advertisements and searches are directed to its rendezvous supernode and the supernode responds back with replies to its own requests. Once the node determines the identity of other peers having the desired file, the data transfer is P2P with the client being able to download the file simultaneously from those peers (downloading different portions of the file from different peers simultaneously).

7.6.2 DHT-based systems

The structured portion of the semi-structured overlays can also be implemented in the form of DHT. The most popular DHT for this purpose has been Kademlia. The initial versions of the Emule file-sharing system used a centralized file directory. The newer versions have support for the Kad Network which is an implementation of the Kademlia DHT.

Another P2P file-sharing system is BitTorrent where each file has an associated *tracker* which keeps track of the peers who have a portion of the file. Thus, there are two levels of resource discovery. First, a peer needs to contact a server to search for a tracker for the file he seeks and then it needs to communicate with the tracker to find out which users have what portions of the file. Each file can have a different tracker, however, in the initial versions there was one tracker node for a file. More recent implementations of BitTorrent have support for sharing tracking of a file by using the Mainline DHT which is based on Kademlia.

7.7 KEYWORD SEARCH USING DHT

DHTs are typically suited for storing and looking up values associated with a single key. Thus, if a record is associated with a unique key, a DHT can be used for a distributed storage and lookup of that record. However, it is not directly amenable to searching for some matching keywords that form part of the key. For example, consider a file with the name 'foo-bar.txt' that is stored in a DHT. In this case, if a user needs to search for any file with the string 'foo' in its name, using the conventional DHT, file 'foo-bar.txt' will not form part of the search results. The reason is that the key corresponding to the entire filename is a random hash and does not have any relation to the text whose hash it represents.

One solution to use DHT for a keyword search is to keep a dictionary of the universe of all keywords. We can compute the hash of each keyword to form a key to be stored in the DHT. The values associated with a key contain the list of files that contain the corresponding keyword. Thus, the search for a keyword reaches the node storing the hash key for that keyword and the returned value contains the names of the files that match that keyword. Second, to find the list of nodes who have that file, the file name itself can be hashed and stored in the DHT with the corresponding values containing the list of nodes having the file.

Another improvement required for keyword-based searching is to allow searches of multiple keywords. For example, a user searching for keys 'foo' and 'bar' should obtain the filename 'foo-bar.txt' in its results but not the filenames 'foo-bas.txt' and 'fop-bar.txt'. One way of implementing such searches over multiple keywords is to issue queries for each keyword separately and then combine the results to filter out those that meet the overall search criteria. Thus, in our example a search for term 'foo' will return both the filenames 'foo-bar.txt' and 'foo-bas.txt' whereas the search for the term 'bar' will return 'foo-bar.txt' and 'fop-bar.txt'. The filter applied to these results would be an intersection of the two resulting sets in case the user is interested in the terms 'foo' and 'bar'. In case the user wants the terms 'foo' or 'bar', the final result would be a union of the two sets. There are several schemes that implement the keyword search mechanism using DHTs with different types of optimization.

7.8 SUMMARY

P2P overlays are characterized by the way the client nodes participate in the network by acting as servers as well as routers. In P2P overlays, the fundamental difference in terms of operations lies in resource discovery. The peer nodes participate in the routing of the resource discovery request and share the request answering the load by storing a portion of the network resource directory. The way these directories are partitioned and the manner in which the requests are directed to a node responsible for the resource in the request content differs in different designs. We have looked at several such methods in the context of unstructured (centralized resource discovery, controlled flooding) and structured overlays (Distributed Hash Tables).

REFERENCES

1. Stoica, I., Morris, R., Liben-Nowell, D., Karger, D., Frans Kaashoek, M., Dabek, F. and Balakrishnan, H. Chord: A scalable peer-to-peer lookup protocol for internet applications, *IEEE/ACM Transactions on Networking* (2002).

2. Karger, D. and Frans Kaashoek, M. Koorde: A simple degree-optimal distributed hash table, *International Workshop on Peer-to-Peer Systems* (2003).

3. Ratnasamy, S., Francis, P., Handley, M., Karp, R. and Shenker, S. A scalable content-addressable network, *ACM SIGCOMM* (2001).

4. Maymounkov, P. and Mazieres, D. Kademlia: A peer-to-peer information system based on the XOR metric, *International Workshop on Peer-to-Peer Systems* (2002).

CHAPTER 8

VoIP OVER INFRASTRUCTURE OVERLAYS

There are two ways in which overlays can be used to deploy VoIP applications. First, the control plane of the VoIP call, i.e. all the auxiliary functionality required to set up the VoIP call (such as user status discovery, codec selection, path setup, etc.) can be off-loaded to the overlay network. Second, the data plane of the VoIP call (the actual voice packet transport) can be offloaded onto an overlay network. The former is done in case there are not enough server resources to keep track of the status of a large user base, i.e. in case there is insufficient infrastructure. The latter arises when one sees benefit in using an indirect VoIP connection between the source and the destination nodes.

For an infrastructure overlay network, it is reasonable to assume that the network has resources to handle the control plane load. Hence, routing the data plane is the more relevant problem there. On the other hand, in a P2P overlay, the lack of large infrastructure implies that the control plane functionality is to be shared among the overlay nodes. While the data plane indirection is not precluded in the P2P setting, it is not as important as managing the control plane functionality.

This chapter deals with efficient handling of the data plane of a VoIP call in an overlay network. The underlying network is assumed to be an infrastructure overlay network so that the issue of nodes joining and leaving the network does not arise. The goal of the chapter is to explain the capabilities that an infrastructure overlay provides in order to enhance the performance of a VoIP call and the user experience.

VoIP application is sensitive to the latency and losses seen by its packets. Hence, deploying VoIP over overlay networks requires that the paths chosen to route VoIP calls over

VoIP: Wireless, P2P and New Enterprise Voice over IP   Samrat Ganguly and Sudeept Bhatnagar
© 2008 John Wiley & Sons, Ltd

the overlay have acceptable delay. Fundamentally, in infrastructure overlay networks, the choice to route VoIP calls on an indirect path through one or more overlay nodes inherently results in paths with a larger number of hops. However, there are several advantages provided by the infrastructure overlay, that there are tangible benefits in using them to transport VoIP calls.

Most of the work described in this chapter has not yet reached the production stage. The insights described in this chapter are gleaned from recently published research from diverse sources.

8.1 INTRODUCTION

VoIP is a latency-sensitive and loss-sensitive application and is therefore susceptible to any delay introduced by the paths over which its packets travel. As we have seen earlier, the routing mechanism in the Internet is not designed with application characteristics in mind. The primary goal of routing is to provide connectivity. Moreover, traditionally, the routing tends to minimize hops between two nodes irrespective of the loss and delay characteristics of the links. Thus, packets belonging to VoIP calls between two nodes receive the same treatment as any other packets between them.

The problem of suboptimality of the routes with respect to VoIP calls is further exacerbated by the routes obtained due to complex interaction of a multitude of routing algorithms that are run within Autonomous Systems (ASes), and the peering relationships between multiple ASes. Effectively, applications susceptible to different network characteristics should be able to create desirable paths when possible. As we already know, one advantage of overlay networks is that it allows applications to route packets through one or more overlay nodes. VoIP applications can leverage overlay networks to route packets so that the *effective* end-to-end delay and loss observed by the endpoints is minimal.

We will explore several features implementable on overlay networks that have been shown to improve the performance of VoIP. We will see when the overlay enhancements improve the quality of VoIP calls and what is the cost of these enhancements. We will see the benefits obtained by applying several techniques in routing VoIP packets overlay networks and how much overhead is incurred in the network in order to achieve those benefits.

8.2 VoIP OVER OVERLAY – GENERIC ARCHITECTURE

Different architectures for routing VoIP calls over overlay networks may differ in specifics, but share a common baseline architecture. We elaborate on this architecture in this section.

In a VoIP overlay there are two types of node in the data plane. The *end nodes* are the client nodes by which the users initiate and receive calls. The *overlay nodes* form the network and participate in routing the VoIP packets. In the control plane, there are one or more *control nodes* which are used to set up the VoIP calls. The three types of node and their roles are summarized in Table 8.1.

An end node starts the VoIP application by registering with the control node, which gives them a list of proximate overlay nodes to which this node can connect. The end node *probes* these overlay nodes to figure out the node to which it has the best quality path. The choice of metric to determine the best quality path may vary among different architectures

Table 8.1   Nodes involved in a VoIP overlay infrastructure.

Node type	Functionalities
End node	Initiate and terminate calls
	Connect to access overlay nodes
	Monitor quality over multiple overlay paths
Overlay node	Monitor path quality to other overlay nodes
	Route VoIP packets in the overlay network
Control node	Provide auxiliary information to set up the VoIP calls

but in general it is likely to be some function of delay and loss seen to different overlay nodes. The chosen overlay node acts as the ingress and egress points for any call from/to this end node. We call this node the access node for the concerned end node.

To initiate a call, the end node contacts its access node to determine if the end node can assist it in improving the quality of the call. If the access node can provide a better quality call, then all packets originating from the end node are sent to the access nodes, which take appropriate actions to determine how to route the packets. Similarly, the other end point of the VoIP call contacts its own access point to determine the route. It is possible that the path taken by packets traversing from end node A to end node B be routed via the overlay whereas the packets on the reverse path follow the direct Internet path. Figure 8.1 depicts a scenario where the VoIP packets from end node A to end node B take a multihop overlay path whereas in the reverse direction the path does not pass any overlay node.

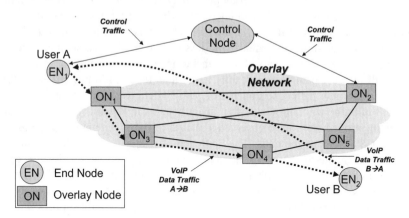

Figure 8.1   The basic VoIP call setup over an overlay network.

8.3   METHODS TO ENHANCE VoIP QUALITY

Several different methods have been proposed in order to enhance the quality of VoIP calls over an overlay network. The fundamental idea is to exploit the path diversity provided by the overlay network and use the extra bandwidth offered by the multiple paths judiciously. Different approaches exploit these avenues differently.

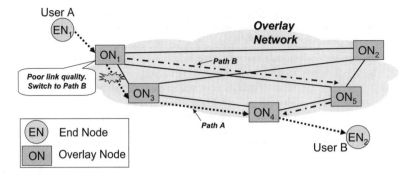

Figure 8.2 Overlay node ON_1 monitors the quality of the path $ON_1 \rightarrow ON_3 \rightarrow ON_4$. When the quality of overlay link $ON_1 \rightarrow ON_3$ becomes poor, the overlay path is switched to $ON_1 \rightarrow ON_5 \rightarrow ON_4$.

8.3.1 Path switching

This approach has been explored in [1]. The fundamental idea is that overlay networks export multiple paths to the application. The paths between multiple overlay nodes have different loss and delay characteristics. Since VoIP is sensitive to loss and latency, judicious choice from among these paths can improve the VoIP quality.

When the VoIP call starts, the overlay network chooses a path that gives it the best quality. However, the traffic dynamics in the Internet are such that the chosen path may not remain the best forever. The other overlay paths may be able to provide better quality for the call. In fact, the quality of paths can fluctuate during the lifetime of a call itself.

To estimate the quality of the paths, the method used is to periodically (around 15 s), probe the paths to determine the latency and loss characteristics. The computed estimate serves as the predictor for the quality of the path during the next time interval. If the expected call quality of the new path is significantly better than the path in use currently, the call is switched to the new path. The concept of path switching is illustrated in Figure 8.2. This process of path-switching yields significant benefits in several settings. The authors of the aforementioned work have observed a margin of improvement of up to 0.87 in the estimated MOS values of the calls. The approach of using the estimated quality in the last time interval as the predictor for the quality in the next interval results in acceptable error. For example, in the case where the margin for improvement was 0.87, the actual benefit obtained using this prediction method was 0.83. In several experiments, this simple prediction model was able to obtain over 60% of the maximum possible improvement (that assumes future knowledge).

8.3.2 Packet buffering

In traditional VoIP setting, the transport method used to carry the VoIP packets is unreliable. The rationale behind not using TCP-like protocols that allow reliable delivery of packets is that the round-trip latency incurred in recovering from packet losses is exorbitant. Since VoIP is latency-critical, the incurred delay in retransmission of these packets makes them meaningless for the receiver. By the time the retransmitted packet arrives at the receiver, it is too late for playback. Conversely, VoIP quality is also affected by the loss. Each packet loss tends to reduce the overall quality of the call. Thus, it is desirable to shield the losses from the application while at the same time the time lag between the two retransmissions of the packet needs to be small enough so that the packet fits in the playout buffer at the receiver.

The presence of overlay nodes at various geographic locations allows VoIP applications deployed over the overlay network to choose one or more nodes on the path to provide temporary buffering. In the case of a packet loss, the egress edge node can then try to fetch the packet from a nearby overlay node rather than going all the way to the other end point and fetch the packets. The goal is to re-fetch the lost packet quickly so that it still has its meaningful place in the playout buffer at the receiver.

This type of approach is used in some research proposals [2]. The overlay network is used to break the long end-to-end path between the end points into multiple but smaller overlay hops. Each overlay node stores a packet in its buffer for 100 ms after it has been transmitted. This allows time for *local recovery* in case the packet does not reach the other end of the overlay link. If a packet is lost on an overlay link, the receiving overlay node detects the loss and asks its upstream overlay node to retransmit it. As shown in Figure 8.3, when overlay node ON_4 detects a packet loss, it asks ON_3 to retransmit it. The packet is retransmitted at most once and then too only if it is expected to arrive at the destination in time for its playout.

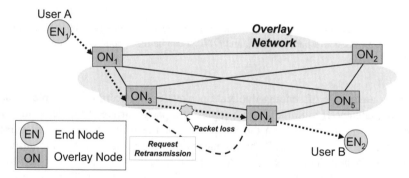

Figure 8.3 Overlay node ON_3 buffers the VoIP packets. If a packet is lost, ON_4 detects it and asks ON_3 to retransmit it.

Another aspect that needs consideration is to design a technique to choose the path to route the call. This path is different from the traditional Internet path as it takes one or more indirections in the form of overlay hops. The routing process is modified to achieve this by changing the metric of an overlay link to represent the expected transmission latency of a packet on that link. This metric includes both the physical latency and that introduced due to expected packet loss and the corresponding retransmission. Computing the shortest path using this metric gives the route that the call should take.

8.3.3 Packet replication

Path switching uses the overlay network path diversity for changing the path of a call in case the original path goes bad. Another avenue of enhancing VoIP quality comes from a different use of the inherent path diversity. Instead of using only one path at a time, multiple copies of a VoIP packet can be sent across the overlay network simultaneously. The idea is that if any one of the copies reaches the destination node, then the losses of the other copies of the packet (at other overlay nodes) become meaningless from the VoIP quality purpose. For example, Figure 8.4 shows a scenario where user A sends VoIP packets through ON_1 which has to route the packets to ON_4. It duplicates each received packet and sends it to ON_3 and ON_5 which forward the packet to ON_4. In case a packet is lost on one of the paths, it may reach ON_4 from the other path. Of course, if the same packet is lost on both paths, it is lost totally.

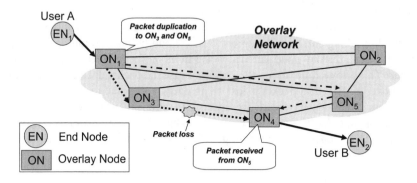

Figure 8.4   Overlay node ON_1 duplicates each packet it receives from user A and sends it to ON_3 and ON_5. If a packet is lost between ON_3 and ON_4, it can be received at ON_4 from ON_5.

Another advantage of sending duplicate packets comes from the possible reduction in perceived latency. If multiple (say n) copies of a packet arrive at the destination node at times $t_1, t_2, \ldots, t_n$, then from the perspective of the receiver, the packet arrives at time $min(t_1, t_2, \ldots, t_n)$. This effectively means that the packet reaches the destination with the latency determined by the fastest of the n paths. An important point to note is that the network characteristics in the Internet fluctuate very fast. Thus, what may be the minimum latency path for one packet might not remain the minimum latency path for the next. However, using packet duplication, this path quality fluctuation is shielded from the application. At a high level, if copies of the packet are sent over all possible sources/destination paths, this effectively leads to path switching on a per-packet basis.

An important point to note here is that in order to use such a strategy, the overlay nodes need to be aware of the intentional duplication taking place. This is not only because the source node needs to replicate the packets, but also for the destination overlay node to filter out duplicate packets when it correctly receives multiple copies of the packet. The filtering needs to be done so that the end-application remains unaware of the strategy deployed inside the network to enhance its experience.

While packet replication has substantial benefits, the cost of replication is high. Replicating each packet n times results in n times the bandwidth usage as compared to the case with no replication. Thus the degree of replication has to be controlled in order to have a reasonable overhead. In practice, such a scheme is only of use in scenarios where there is sufficient spare bandwidth available and there is substantial VoIP quality gain by using this scheme. For example, if by using one extra copy of a packet, the R-factor of the VoIP call increases from 70 to 72, it may be overkill to use replication whereas if the R-factor improves from 50 to 72, it may be well worth the extra bandwidth.

8.3.4 Coding

The last section clearly highlights the benefit of using multiple copies of a packet through the overlay network to reduce the effective latency and to conceal packet loss. The only problem with this approach is the associated bandwidth cost. This problem can be mitigated by appropriate use of erasure coding methods.

Erasure coding is a method of encoding n bits of information in such a way that loss of any k bits out of n can be concealed. In other words, if we flip any k bits from the original n bits, an erasure coded block of the n bits can automatically identify and recover the flipped bits. The concept of erasure coding can be used to encode more than one packet at a time. Thus, rather than having duplicates of each packet, an alternate is to have k extra packets to encode a group of n packets. If the receiver receives any n out of these $n + k$ packets, all n packets can be recovered. This scheme retains the benefits of packet replication because any k losses can be concealed. Also, the first n packets received are the only ones relevant and any packets from this group arriving subsequently are discarded. This has the impact of reducing the perceived latency for the group of n packets. Furthermore, the overhead of using this scheme is less than duplication if $k < n$.

However, use of coding requires extra computation at the sender and the receiver. The sender has to compute the content of the extra k packets given a block of n packets. The receiver needs to decode the received n packets to figure out the original n packets. Due to the real-time nature of VoIP, the encoding and decoding process should not be complex and time consuming. For example, if the encoding/decoding process takes more than 200 ms, then the packets will reach the user substantially delayed resulting in a poor voice quality. Along with the use of simple codes, the number of packets that can be treated as part of one coding block has to be low. The reason is that accumulating packets to allow coding also delays them. The choice of a coding block size depends on the time consumed by the encoding process, the network latency and the decoding process.

A simple yet meaningful code useful for VoIP is parity coding. The parity coding technique adds 1 bit to a block of n bits. The parity bit is set by counting the number of 1s in the n bits being encoded. If the convention is to have odd(even) parity, then the total number of 1s in the original n bits along with the added parity bit is odd (even). Therefore, in case of odd parity if the number of 1s in the n-bit block is even, the parity bit is set to 1 and otherwise it is set to 0. Now, the value of any 1 bit lost among these $n + 1$ bits can be computed using the known parity. This process is applied at the packet level where we have n packets being encoded along with a parity packet. In this case, the ith bit in the parity packet is the parity bit for the block of n bits comprising of the ith bits of the n packets being encoded. Thus, if any one of these $n + 1$ packets is lost in the system, it can be reconstructed using the remaining packets. However, if two or more packets are lost, then the contents of the these lost packets cannot be reconstructed. However, the packets correctly received (except the parity packet), can still be played out normally.

Another method of encoding that can used is called Multiple Description Coding (MDC). Here different descriptions of the voice stream are sent over different network paths. If all descriptions are not received by the playout time, the available descriptions are combined into a somewhat lower quality voice signal. The more the number of descriptions received, the better is the quality of the reproduced signal. In this approach, it is tough to establish the quality gains of using MDC. The reason is that in conventional metrics, the loss and delay along with the knowledge of the codec used gives an estimate of the VoIP quality. However, with MDC, the loss of packets (descriptions) implies a lower quality signal being reproduced from the received descriptions. Thus, while a potentially lower quality signal can be reproduced from the received descriptions, how the quality of received voice can be quantified is not clear.

8.4   ESTIMATING NETWORK QUALITY

Overlay nodes establish connections to each other and keep an updated picture of the quality of links in the overlay network. Similarly the user nodes need to estimate the quality of the paths to the overlay nodes. To have an estimate of the quality of links and the quality of paths, the overlay nodes need to monitor various parameters of the overlay links. These parameters need to be transformed into a quality of link estimate. Further, from the estimate of link qualities, the overlay nodes need to find out the quality of the paths in the network. Different versions of the measurement techniques are used in the referred works. We discuss generic probing methodology and identify the crux of the techniques used. This description is based on the model used in [3].

We now look at the details of how end nodes and overlay nodes assess the quality of the overlay link between them. In most cases of using VoIP, the R-factor serves as a reasonable metric of the link quality. As we have seen earlier, R-factor is a function of delay, the network-loss rate, the jitter-loss rate and cluster-factor. The definitions of these parameters for the purpose of this measurement are given below:

- **Network Delay (d):** If t_s is the time at which the packet was sent over the link and t_r is the time at which it was received at the destination, then $d = t_r - t_s$. This is the network delay faced by the packet while traversing the path between the source node and the destination node. It does not contain the delay incurred due to the voice processing in the codec and the delay introduced by the jitter-buffer.

- **Network Loss Probability (n):** The network loss probability is the probability that a packet was sent but did not reach the destination. Hence, $n = P$ (packet is dropped) and is computed as $(\#packets\ sent - \#packets\ received)/\#packets\ sent$.

- **Jitter Loss Probability (j):** The jitter loss probability is the probability that the packet is dropped due to late arrival, i.e. the packet reaches the destination over the network but has to be dropped because it arrived past its deadline and cannot be played out. Since jitter buffer adds a smoothing delay T_s before it plays out the packets, if a packet arrives more than T_s later than the delay incurred by the *first* packet of the current burst of packets, it is deemed to have missed its deadline. Thus, $j = P((d - d_f) > T_s)$ where d_f is the network delay incurred by the first packet in the current burst of packets.

- **Cluster Factor (c):** The cluster factor represents the conditional probability that a packet is lost when the packet prior to it was also lost. $P(p_{i+1}$ is lost $|p_i$ is lost), where p_i indicates the ith packet in the burst. The importance of cluster factor arises when certain codecs such as G.711 are used. It has been observed that for the same loss ratio, if the losses were random the R-factor observed was higher than that computed when the same amount of packets were lost but the losses were bursty. Thus, if such codecs are used in the VoIP call, estimating the quality requires an idea of the burstiness of the losses. This estimate of burstiness is provided by the cluster factor.

The goal of the link quality estimation technique is to deduce the values of these parameters accurately. To this end, each node has two options: if it has an ongoing VoIP call to another node, it monitors the packets in the stream; if not, then it has to simulate a VoIP call by sending out a stream of packets to the other endpoint of the overlay link to be measured. In either case, we can think of the stream of packets as *probe traffic*.

8.4.1 Probe traffic

Each end node and overlay node periodically probes the quality of the link between the associated nodes. Additionally, the nodes can test the quality of links to a subset of other overlay nodes so that they may be able to identify if a better link exists for them to use to route certain traffic. After all, without probing, there is no way for a node to tell if the quality of a potential overlay link is good or bad. On the other hand, being overly aggressive and consistently sending probe traffic to all nodes in the network makes the system unscalable since each of the N nodes probing all other $N - 1$ nodes results in $N(N - 1)$ probes which makes the overhead $O(N^2)$. Thus, each node chooses to probe only a subset of other nodes. The choice of the subset can vary. For example, a node may probe the quality to only its nearby nodes – those having a network latency less than a set threshold value.

To probe the quality of link to a prospective node, each node (end node or overlay node), sends out a stream of dummy VoIP packets. The packets are tagged with the time-stamps at the instant they are sent out (t_s) in the network. The receiving node receives the packets and stamps them with the time (t_r) at which they were received. The duration of the probe stream should be small enough so as not to incur too much overhead. At the same time, the stream should have sufficient number of packets so that the gathered statistics are meaningful. For example, authors in [3] used 15 s bursts for their probe streams. With each packet encoding 15 ms worth of voice signal, this amounted to 1000 packets. The boundary of the burst is marked using special packets so that the receiving node can estimate the required parameters knowing the span of the burst. The receiving node resets its estimate values on receiving the burst-start notification and starts recording the statistics for the entire burst. After receiving the burst-end notification packet and completing its calculation of statistics, these values are returned back to the probe source node. In case of a centralized control node, these values are reported to the control node as well. The calculation of the statistics from this burst is described next.

8.4.1.1 *Network delay (d)*   Estimating delay seems like a simple method where the difference between stamped send-time t_s and its receive time t_r is calculated. However, this simple method is insufficient because there are two deeper problems that need to be overcome. First, the delay incurred by different packets may vary because of the variation in the Internet cross-traffic they encounter while travelling over the multiple hops comprising the overlay hop. Second, the clocks used to stamp the t_s and t_r values are at different nodes and are unlikely to be synchronized. In an extreme case, the clocks may be so skewed that the value of t_r may become less than t_s. Clearly if these values are used, the packet would seem to have arrived before it was even sent. Thus, estimating the network delay is not as simple as it seems at an initial glance.

A reasonable estimate of the network delay can be derived using the Round Trip Time (RTT) between the two nodes. This requires bi-directional communication between the nodes. The probe-receiving node also sends traffic to the sending node with similar time stamps recorded in the packets. Using the receive and send time-stamps, the skew in the two clocks can be eliminated. Consider that the accurate time at the instance when node A sends out a packet to node B be T. Let the clock skew at node A be s_A so that it stamps the outgoing packet with a time value of $T + s_A$. The packet incurs the network delay d reaching node B at the accurate time $T + d$. If node B has a skew of s_B from the actual time, it will estimate the receiving time as $T + d + s_B$. Hence the delay estimated by B on receiving this packet using the time stamps is $(T + d + s_B) - (T + s_A) = d + s_B - s_A$. Hence the respective clock skews as the two nodes distort the estimated delay. However, if B sends a packet to A, by symmetry of the argument and assuming the same network delay d on the path from B to A, the latency estimated by A would be $d + s_A - s_B$. If A knows the latency estimate of B, it can add its own latency estimate to it and obtain $(d + s_B - s_A) + (d + s_A - s_B) = 2d$. This new estimate eliminates the clock skew from the equation altogether and taking half of this value as the estimate of the network latency is a feasible solution. However, one caveat is in order – the paths on the Internet are not symmetric. Thus, it may be possible that the network delay incurred on the path from A to B be different from that observed on the reverse path. Hence, while using this technique of halving the RTT to estimate the one-way latency eliminates the problem of clock skew, it still does not give the best solution. Furthermore, the estimated delays for different packets may vary because they may face a different amount of instantaneous cross-traffic and queues en route. Thus, the estimated network delay has to be averaged over several packets.

It must be noted that the estimation of network delay can be done in several ways. Several refinements of the aforementioned process are possible. For example, if the nodes are in the control of a single administrator, the nodes can periodically use the globally available Simple Network Time Protocol (SNTP) servers to synchronize their clocks to the reference servers. This would virtually eliminate the skew and obviate the need to estimating the one-way latency using RTTs. Furthermore, statistical tools can be used to refine the estimate of the one-way delays using only time stamp information.

8.4.1.2 Link jitter loss (j)
To estimate the jitter loss in a link, the receiving node has to calculate the delay for each received packet *relative to* the first received packet in the probe traffic. Suppose that the first packet is received at time t_{r_1}, and the jitter buffer playout threshold is T_j, then if the delay incurred by *any packet* is T_j or more compared to the delay of the first packet, that packet is termed to have a jitter loss.

The actual value of T_j depends on the size of the jitter buffer and the codec used. For example, suppose we are using G.711 codec with each frame encoding 15 ms worth of data. If we have a static jitter buffer storing ten frames worth of data and the first frame is played back after six frames worth of data is filled in, then the amount of jitter buffer delay T_j allowed is 15 ms $\times 6 = 90$ ms. Thus, any packet whose delay is 90 ms or more compared to the delay incurred by the first packet in the current burst, then the new packet is too late for its playout and will be dropped as a jitter loss. On the other hand, a packet arriving within a delay jitter of 90 ms compared to the first packet is buffered and will be successfully played out. At the end of the probe burst, the receiving node calculates the jitter loss as the fraction of the packets in the burst that arrived safely at the jitter buffer but were dropped because of their late arrival.

8.4.1.3 Link network loss (n)
The estimation of the loss sustained at the monitored overlay link is straightforward. Here the receiving node keeps track of the number of packets received in the system and estimates the loss using the unique IDs of the packets. The number of packets lost is given by the set of IDs not received. The ratio of such lost

packets and the total number of packets transmitted gives us the estimate of the network loss for the link.

8.4.1.4 *Link cluster factor (c)* Estimation of the cluster factor for the monitored link is similar to the estimation of the network loss. Here, instead of counting the number of losses, the receiving node counts the number of losses such that the packet with the immediately preceding ID is also lost. This is an approximation of the degree of clustered losses going on in the network. It is possible to have more complicated measures of clustered losses where, rather than just looking at the probability of consecutive losses, we consider a sequence of losses as a single burst of losses rather than a bunch of pair-wise losses. However, practical experience has shown that using such a simple approach to loss clustering has substantial benefits in itself.

8.4.2 Estimating path quality

In the above section we discussed how network delay, network loss, jitter buffer loss and clustering loss is estimated. These can directly be used to estimate the R-factor of any overlay link. However, the path for a call through the overlay network can comprise multiple hops. This entails estimating the R-factor of an entire path rather than that of just individual links. This estimation is not straightforward since the R-factor is *not additive* as it depends on the packet loss which is not additive. Thus, we cannot directly add the R-factors of multiple overlay links to obtain the R-factor of the path. Furthermore, the R-factor is not a bottleneck metric since it depends on delay, which is additive. Hence, we cannot take a minimum of the R-factors of the constituent overlay links to estimate the R-factor of the path. Thus, to compute the R-factor of the path, we need to estimate the delay, network loss, jitter loss and cluster factor for the entire path from the computed values of the individual overlay links. If we try to measure these values for all possible multi-hop paths in the overlay network, the computation cost becomes prohibitive. The number of overlay paths that can be formed in an overlay network with n nodes is $O(2^n)$ if each overlay node can connect to any other overlay node. Even if we restrict the size of the potential neighbor set of each overlay node to at most D, the number of overlay paths is $O(2^D)$. Thus, the only feasible way of estimating the path quality parameters is to use the values determined for the overlay links and combine the values of the constituent overlay links. How these values are combined is described next.

8.4.2.1 *Path delay* Delay is an additive quantity. Thus the delay faced by a packet traversing a path comprising two links with delays d_1 and d_2 respectively is given by the sum of the delays of the individual links *Path Delay* $= d_1 + d_2$. For a k link path with delays $d_1, d_2, \ldots, d_k$, the path delay is given by $\sum_{i=1}^{k} d_k$.

8.4.2.2 *Path network loss* Network error is a multiplicative quantity. Consider a sequence of packets that passes through two overlay links with error probabilities n_1 and n_2, respectively. Each packet passes through the first link with probability $(1 - n_1)$. Of the packets that pass through the first link, only a fraction given by $(1 - n_2)$ make it through the second link. Thus, total error induced by this pair of links is given by $1 - ((1 - n_1)(1 - n_2))$. Extending this relation for a path with k links with loss probabilities $n_1, n_2, \ldots, n_k$, the cumulative loss probability of the path is given by $1 - \prod_{i=i}^{k} (1 - n_i)$.

8.4.2.3 Path jitter loss Estimating jitter loss for a multi-hop path from the jitter-losses of individual links is more involved than the computation of delay and loss. To compute the path jitter loss, an estimate of the jitter loss at each of the links is needed and combined into a single jitter loss metric.

Consider a scenario where the size of the jitter buffer is static where a packet is lost if it has a delay that is greater than the sum of the delay of the first packet and the jitter buffer threshold. In this case, the probability of loss depends on the delay distribution of the packet. Suppose the delay distribution of packets on a given link l is $d_l(t)$, then the jitter-loss probability j_l on link l can be computed as $j_l = \int_{d_{1_f}+T}^{\infty} p_l(t)\,dt$ where d_{1_f} is the delay faced by the first packet and T the jitter-buffer threshold. This value signifies the sum of probabilities of having a delay more than the jitter-buffer threshold.

The reason why combining the jitter-buffer loss probabilities is tough is because a packet can be lost in jitter buffer if the *sum* of the delay it incurs over all the links of the path is over the sum of the delay faced by the first packet and the tolerance threshold of the jitter buffer. Assuming that the delay distributions of packets on two links are independent, the combined delay distribution of packets traversing the two links 1 and 2 sequentially is given by $p_{12}(t) = p_1(t) \otimes p_2(t)$ where $\otimes$ is the convolution operator. The jitter loss j_{12} of the two links is $j^{12} = \int_{d_{1_f}+d_{2_f}+T}^{\infty} p_{12}(t)\,dt$. This quantity represents the total probability of packets incurring a cumulative delay at least T units more than the cumulative delay $d_{1_f} + d_{2_f}$ faced by the first packet of the burst over the two links. Thus, we need to know the delay distributions of links along with the delay of the first packet to find out the jitter buffer loss.

Passing around the delay distributions themselves results in a significant overhead. A simpler strategy is needed to alleviate this limitation. Experiments have suggested that adding the jitter-loss probabilities gives a reasonable estimate of the complete path jitter buffer.

8.4.2.4 Path cluster factor For a set of packets traversing two links 1 and 2, the cluster factor for the complete path can be computed as a sum weighted by the loss rate. Thus $c_{12} = (n_1 \times c_1 + n_2 \times c_2)/n_{12}$. Note that the cluster factor is only computed for the packets that are lost. Hence the above formulation indicates the clustered losses that occur on either link 1 or link 2, given the probability of loss over the path comprising the two links. This estimate simply takes the total probability of clustered losses on link 1 or link 2 and computes the fraction of times this event occurs among all losses occurring on either link. This estimate serves as a good approximation to the actual cluster factor of the path comprising the two links.

8.5 ROUTE COMPUTATION

Having estimated the delay and loss metrics for individual links, using these techniques, it is possible to compute the delay and loss characteristics of any path in the overlay network. This implies that the R-factor that any path can provide can be computed and the qualities of paths can be compared on a common ground. The call routing algorithm needs to estimate the quality of the paths and then decide to route the call accordingly.

The route computation process can be centralized or distributed. In a centralized implementation, the nodes report the locally estimated values of all overlay links to the control node. The control node uses the metric-combination techniques described above to estimate qualities of the sub-paths and chooses the best path based on its estimate. The route computation process can be distributed, also. This can be done using standard routing algorithms except that we have multiple metrics to propagate and the combination of metrics requires special attention.

An example routing implementation could be the well-known Distance Vector technique. In this setting each overlay measures the quality of its links and reports the measured values to all other nodes. After the message exchange, each node figures out the best path through the overlay for any given source/destination pair. If a node finds that a better route is available for some end node through the overlay network, then the routing information is updated and propagated. The route computation is invoked periodically.

8.6   PERCEIVED ENHANCEMENT OF VoIP QUALITY

While the above techniques work in theory, several efforts have tried to figure out in the real world, how well these methods fit. The goal is to measure the perceived improvement in the VoIP quality in the presence of so many diverse factors affecting its performance.

Different tools such as simulation and emulation have been used to measure the performance of VoIP in overlay networks when the described techniques are used. We highlight the potential benefits of the use of overlay networks that are visible in some of the performance results shown in [3] where the experiments were run on the PlanetLab test-bed. In all, 31 nodes located in different places in the world were used as the test machines. Of these machines, 11 were chosen to act as overlay nodes depending on the number of paths on which these nodes helped reduce the latency. The chosen overlay nodes helped in reducing the end-to-end latency of at least three pairs of end nodes. The VoIP traffic generated travelled over this network and interacted with other normal Internet traffic.

The experiment used the G.711 codec which is a high bandwidth codec (64 Kbps). The sampling interval for the codec is 15 ms resulting in a 132-byte RTP packet and a 160-byte IP packet. The total bandwidth consumed by one bi-directional call is approximately 170 Kbps. The packets emulating a VoIP call are generated in bursts of 1000 packets (equivalent to 15 s of talk). Each call consists of multiple such bursts.

In this simple setting, several benefits of using the overlay network were observed. We classify these benefits into sub-categories:

- *Overlay Routing:* On an average more than 40% of calls saw improvement when they were routed using the overlay rather than directly connecting the endpoints over the basic Internet path. The average R-factor of these calls increased by a value of 26. Furthermore, simply by routing over the overlay, 11% of calls improved their quality to an acceptable level and a further 11% were improved from an acceptable to a high level.

- *Path Diversity:* The use of multiple paths to reach a destination is proposed as a way of improving the voice quality. Thus, two calls between the same pair of overlay nodes may be routed along different overlay paths rather than following the same path as would be the case in the Internet. In the experiments, the number of paths used for routing the calls was increased from one to four. The quality of the calls is monitored to have an estimate of the benefit of path diversity.

   The observed improvement in the R-factor increases with the number of paths but the marginal improvement reduces with the number of paths. The quality of the number

Table 8.2   VoIP enhancement using path diversity.

# Paths	Average increase in R-factor	Decrease in low-quality bursts (%)
2	4	11
3	11	21
4	12	22

Table 8.3   VoIP enhancement using packet replication.

# Paths	Average increase in R-factor	Decrease in low-quality bursts (%)	Bandwidth used (Kbps)
2	18	27	340
3	34	48	510
4	44	64	680

of individual bursts of packets that simulate talk time for one user went up. The results of this set of experiments are summarized in Table 8.2.

- *Packet Replication:* Packet replication utilizes additional available bandwidth by sending out multiple copies of each packet in a VoIP call. The intention is to reduce the effective loss probability as the probability that at least one copy of the packet will reach the destination is higher than when there is only one copy. The experiments test the improvement in the R-factor and the reduction in low-quality talk bursts if two, three or four copies of the packet are transmitted.

The observed benefits of replication are shown in Table 8.3. The advantages of duplication can clearly be seen. While in the best case with four copies, the R-factor jumped up by 44 and the low-quality talk spurts reduced by 64%, this comes at a cost of increased bandwidth overhead which multiplies with the number of copies.

- *Parity Coding:* Sending duplicate packets improves the voice quality but consumes a large amount of additional bandwidth. Parity codes can reduce the bandwidth overhead yet retain some benefits of the loss-resilience provided by duplication. This is attained by adding one parity packet to a block of k packets. If any k out of these $k + 1$ packets are received, the original k packets can be reconstructed. To test the benefits of parity coding, experiments were run with $k = 2$. The R-factor increased by an average of 16 with reduction in low-quality bursts by 27%. This is achieved while using only 255 Kbps bandwidth (an overhead of 85 Kbps only).

8.7   SUMMARY

Overlay networks can be used to enhance the quality of VoIP calls. A VoIP call enters the overlay network from an ingress overlay node and leaves from an egress overlay node. The overlay network can take various forwarding decisions for the VoIP packets. These include switching paths dynamically when the original path's quality deteriorates, local recovery to conceal losses, sending replicating packets over diverse paths to reduce effective packet losses, and using coding over a set of packets to provide the benefits of replication at a lower cost. All these tasks require estimation of overlay link properties from simple measurements and need to compute routes based on the end-to-end path quality estimation. The techniques have been shown to be very effective in several experiments.

REFERENCES

1. Tao, S., Xu, K., Estepa, A., Fei, T., Gao, L., Guerin, R., Kurose, J., Towsley, D. and Zhang, Z. Improving VoIP quality through path switching, *Proceedings of IEEE Infocom* (2004).

2. Amir, Y., Danilov, C., Goose, S., Hedqvist, D. and Terzis, A. 1-800-OVERLAYS: Using overlay networks to improve VoIP quality, *Proceedings of ACM NOSSDAV* (2005).

3. Rajendran, R., Ganguly, S., Izmailov, R. and Rubenstein, D. Performance optimization of VoIP using an overlay network, *NEC Laboratories Technical Report TR2006-08, (http://www.ee.columbia.edu/~kumar/papers/tr06-08.pdf)* (2006).

CHAPTER 9

VoIP OVER P2P

VoIP over infrastructure primarily deals with the packet routing, coding and forwarding in the data plane of the call. In a P2P overlay, while these techniques are applicable, the more important task is the management of the VoIP control plane in the network. Infrastructure overlay networks would presumably have sufficient auxiliary resources to manage the control plane of the voice calls. However, this is not true in the case of P2P overlays where the overlay nodes are essentially the typical low-end home computers instead of powerful servers. Thus, in a P2P network the task of control plane management also has to be shared among the nodes. Furthermore, if the nodes have a high degree of churn, then routing calls through these nodes is not desirable because if the node leaves, then the call will have to be re-routed. Of course, using more stable overlay nodes for call relaying purposes would definitely have advantages like that in an infrastructure overlay, such call routing becoming a secondary consideration in the P2P setting.

This chapter deals with the control plane issues that an overlay network faces when a VoIP service is deployed on top of it. The idea is to establish the baseline services that are required in a P2P overlay to support VoIP. Different P2P VoIP overlays may implement one or more of these features in their own specific ways. For an insight into a real P2P VoIP overlay, we will take a look at the working of the popular Skype network.

9.1 VoIP OVER P2P OVERLAY – GENERIC ARCHITECTURE

The architecture for supporting VoIP calls using a P2P overlay is different from that in an infrastructure overlay. The architectural differences are twofold:

VoIP: Wireless, P2P and New Enterprise Voice over IP   Samrat Ganguly and Sudeept Bhatnagar
© 2008 John Wiley & Sons, Ltd

- There is no central control node that stores the status information of all nodes. The information has to be distributed among various nodes.

- The VoIP packets are likely to be routed directly between the two end nodes. The overlay nodes may be used in certain cases, but it is unlikely to be to enhance the VoIP quality as is the case when the data plane is routed over the overlay network.

These differences lead to an architecture where the control node functionality is distributed among a subset of the overlay nodes whereas the data plane traffic is direct. Different P2P networks can choose to distribute the control plane management responsibility among nodes differently.

The generic architecture is shown in Figure 9.1. When User A connects to the network, it registers with an overlay node using a control message. The registration process would update the status of the user as recorded in the overlay network. Similarly, User B can register with the overlay network when it comes online. When A wants to talk to B, it needs to figure out at least two things: whether B is online; and what is the IP address of the machine on which B is logged in. This is the type of information that the registration process stores in the overlay network. As discussed in Chapter 7, there are several techniques by which this information can be stored in the overlay. The exact method by which this information is stored in the network and how it is searched and retrieved from the network may vary in different P2P networks.

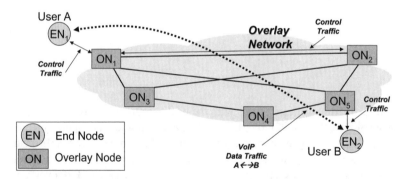

Figure 9.1   The generic architecture for supporting VoIP over a P2P overlay.

Once A finds out that B is online and comes to know of its IP address, it can initiate a call to B directly. The data plane may or may not be routed through the overlay network. In fact, in practice, the routing of data plane through the overlay network is not primarily for the purpose of quality enhancement. The overlay nodes may be used purely for rendezvous and relaying purposes when either A or B or both are behind NAT devices. We shall study such scenarios in detail later.

9.2   VoIP ISSUES IN P2P OVERLAY

When using P2P overlay as the basic network for managing the VoIP calls, several practical issues arise. These issues relate to the fundamental properties of the P2P overlays. We broadly classify these issues into two categories based on whether they arise due to the architecture of P2P overlay where end-nodes are privy to sensitive information or due to the properties of the created overlay network.

9.2.1 Architectural issues

The fundamental architectural issue that arises in the P2P setting is that the end nodes form part of the infrastructure. It is easy to envision scenarios where malicious nodes join the overlay and become part of the network. Since the P2P architecture allows storage of potentially sensitive user information in the network, the malicious nodes can acquire information which they could use in a negative manner.

There are two types of process where this kind of information arises:

- *Authentication:* At the start of an end node's association with the P2P overlay network, the corresponding user needs to be authenticated. In a P2P setting, this task is the most challenging because it is end-hosts that are acting as the overlay network. The requirement is to keep the user's password information secret and yet allow the network to verify the user's identity. This is the single most critical functionality that the network needs to provide. The reason is that if the password is broken, malicious use of the user's account is possible. The problem is exacerbated in case the user is paying for the service.

- *Presence Information:* Maintaining the user privacy is also related to the security risk posed by the P2P architecture where the end-nodes store user information. The user stores its relevant information in the network. This information includes the current IP address, the presence status which at the very least includes the information regarding the availability of the user and the types of codec that the user can support. If a malicious user has access to this information, he will have details regarding the user that were not meant for him.

Both these issues require that the control plane traffic be encrypted and digitally signed end-to-end. Thus, if a user's password is to be verified in the overlay network, the network should ensure that the password is never sent out in a clear text and is not a constant string (that encodes the password). Similarly, to guard against a malicious user changing the content of the user presence information (for example, changing the IP address to redirect the communication to another machine), any updates to a user's information should be in a message digitally signed by the user. We will look into some of these details later.

9.2.2 Network issues

While the previous issues arise from the specific usage of P2P nodes as the information stores, another set of problems arises due to the scale and diversity of the overlay network itself. The issues that fall in to this category are as follows:

- *Information Availability:* The P2P overlay nodes are the typical home computers. The users owning the computers may join the P2P network, make the VoIP calls in which they are interested and then disconnect from the network. This constant joining and leaving of users is a norm in the P2P world in contrast to the infrastructure overlay nodes whose sole purpose is to support and enhance the VoIP service. Due to this user churn, the VoIP service deployed in the network needs to be carefully designed for information availability. Any user information that is stored in the network may become unavailable if the node on which it is stored leaves the network. Naturally any such information has to be replicated and that too onto nodes which are expected to last longer in the network.

Furthermore, if information is available on a single node in the network, the node may not leave the overlay network but still may not be reachable. This can happen due to disruptions in the underlying physical network, leading to temporary instability in the network paths. A node can also remain reachable but it may not be able to provide access to the stored information. This scenario can occur in case the node is facing denial of service attack.

Several types of network disruption can cause problems in the P2P overlay network. These issues need to be addressed in order to deliver VoIP over P2P overlays.

- *Ease of Search:* Ensuring information availability allows the nodes to *store* information in such a manner that it remains readily accessible when a user wants to find another's status. However, in a large overlay network, locating a node which contains the desired information in itself is a tough task. A user interested in making a call to another would expect that the status of the callee (availability and IP address in particular) be known at the earliest. Thus, if the searching process is slow, the control plane functionality provided by the P2P overlay may be rendered useless in practice.

As we discussed earlier, there are mechanisms to achieve fast searching in P2P overlays. This can be achieved by using structured or semi-structured overlays.

Having identified the issues that need to be addressed in order to provide a good quality VoIP service over a P2P overlay, we next look at a specific solution that has gained immense popularity – Skype. We shall gain several insights from the design choices used in the Skype P2P network and will understand the real-world observations regarding the performance of the Skype network.

9.3   CASE STUDY: SKYPE

Perhaps the most famous P2P overlay application supporting VoIP is Skype. Ever since it was launched in 2002, Skype became the VoIP application of choice in the Internet. The widespread adoption of Skype emanates from two root causes:

1. From the onset, Skype allowed users to have high-quality PC-to-PC VoIP calls for free. While there were existing players providing PC-to-PC VoIP calls, the general observations regarding the quality of these softwares were not very encouraging. Skype received rave reviews from the users from the very beginning.

2. Skype was built by the designers of the Kazaa P2P network. The team had access to a huge user-base in the FastTrack network that supported Kazaa. The quality of VoIP provided by Skype led to quick adoption among the Kazaa user population. Furthermore, with the time-tested FastTrack P2P technology that enabled storing and searching of records (file-names in the P2P setting), the control plane tasks became relatively simpler. With a good codec at their behest, the team could focus on providing a good-quality call without having to worry about the control plane details. Furthermore, the need for infrastructure was obviated by the large fast-track user-base.

We use Skype as a representative VoIP network for an understanding of how it solves the aforementioned issues arising in VoIP in the P2P overlays. We also look at the results of a few studies undertaken to understand the behavior of the Skype network. Since the Skype protocol is not open, the understanding that follows in this section is based on the studies that have tried empirically to determine the operation modality and the achieved performance.

9.3.1 Skype architecture

Skype is based on the FastTrack network technology that underlies the Kazaa network. As discussed in Chapter 7, only a subset of the protocols used are known publicly through studies like that in [1].

There are two types of node in the Skype network:

- *Skype Client (SC):* A Skype Client node is an ordinary end node in the network that runs the Skype software purely as a client.

- *Super Node (SN):* A Super Node in the Skype network is an overlay node that helps in storing and searching user information along with relaying of calls to handle NATs. Any node that is not behind a NAT and has sufficient CPU power and network bandwidth can automatically become a SN. The nodes do not have control of whether or not they become SNs.

The SC stores a list of SNs to which it can connect. This list is stored in a repository called the *Host Cache* (HC). Among other things, the HC contains the IP addresses/ports of a few SNs to which the SC can connect.

Skype uses the overlay network of SNs to perform the majority of the control plane tasks such as user directory storage and lookup. This overlay network is called the 'Supernode P2P Architecture'. Each SC connects to one SN and each SN can have multiple SCs connected to it. In order to ensure a high level of security and availability, each user's authentication information is stored in a Login server that forms the sole infrastructure component in the Skype network. A conceptual representation of the Skype architecture is shown in Figure 9.2. There are a bunch of SNs (labelled SN_1 through SN_5) each acting as the access point for several SCs. Each SC stores the IP addresses and port numbers of several SNs in its HC. As an example, the HC at SC_1 stores the IP addresses and ports of two SNs. Of these SNs, SC can connect to anyone (in the figure it connects to SN_1).

Like any P2P network supporting VoIP, Skype has to address the aforementioned issues. While the network-related issues are a performance concern, the information-confidentiality issues that arise due to the architecture are a show-stopper if they are not addressed. In order to address the confidentiality issues in a systematic manner, Skype has defined a set of security guidelines that the design meets. The stated security policies of Skype are:

- All the usernames must be unique so that no users can impersonate each other. The use of a central login server helps to ensure this.

- Users who wish to use Skype must first authenticate themselves by presenting their credentials, including the assigned username and the password.

- When a user wishes to communicate with another, the two must present each other with credentials that they can verify to assert the other's identity.

- Each message exchanged in a Skype VoIP session is encrypted end-to-end. Any node intending to eavesdrop on the communication is only going to see encrypted traffic. Only the node to whom the traffic is destined can decipher the packets.

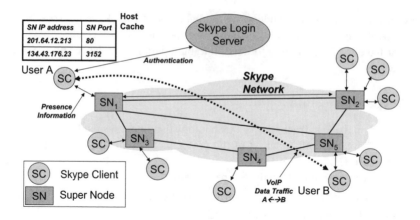

Figure 9.2   Architecture of Skype P2P overlay.

All Skype operations are directed towards ensuring these policies. Additionally, the Skype code is closed-source so that it is tough to find security loopholes. In fact, it is highly obfuscated to remain shielded from attempts to debug the machine code with several sophisticated traps to foil attempts to reverse engineer the code. Having noted the impact of the possible concerns with any P2P architecture that allows nodes to store sensitive information, perhaps the extent of protection provided in the Skype code base is justified.

We next describe the information regarding the Skype operation that is known publicly.

9.3.2   Skype operation

We now describe how various operations are performed in Skype. The details are at a high level and are intended to give insights into the design choices used in the process.

9.3.2.1   *Installation and configuration*   The Skype client installer can be downloaded from the Skype website. While the most popular operating system used by clients running Skype is Windows, Skype is also available in binary form for Mac and Linux. On installation, Skype stores its main configuration and user-specific information on the local machine in two XML files. These files are called *shared.xml* and *config.xml*. They are stored in different locations depending on the operating system. For example, in Windows the shared.xml file is stored in the `C:\Documents and Settings\windows_user_name\Application Data\Skype` directory where `windows_user_name` is the Windows login name of the user. The config.xml file is specific to the Skype user and is stored in the directory `C:\Documents and Settings\windows_user_name\Application Data\Skype\skype_login_name` directory.

The file shared.xml stores information for the SC that is used by all users who may use that client. The most prominent information that shared.xml stores is a set of IP addresses and ports of the SNs that are part of the network. As mentioned earlier, this storage unit is the HC. The client keeps information about recently reachable SNs in the HC and keeps updating this information periodically. Other information that is stored in this file includes details about the host machine such as the bandwidth, whether it is behind a NAT, whether the node is behind a firewall, the port on which Skype listens to the incoming messages and the types of input/output device to use on the host machine.

The content of config.xml represents the user-specific parameters for each Skype user. Of these, the most prominent one is the list of contacts for the user. For each contact, an XML tag with the name of the contact is created and a 4-byte ID is stored inside the tag. As observed in Skype version 3.5, this information is stored in the tag hierarchy *<Central Storage>* → *<SyncSet>* → *<u>* → *<contact_name>*. Thus for two contacts CN1 and CN2, this portion of the file looks like:

```
<CentralStorage>
    <SyncSet>
        <u>
            <CN1>f3359d4b:2</CN2>
            <CN2>7a43fac8:2</CN2>
        </u>
    </SyncSet>
</CentralStorage>
```

In addition to storing the list of contacts for the Skype user, config.xml file also stores a great deal of other configuration variables such as what messages to display and what type of sounds to play to alert the user to events such as new call arrival, Voice Mail setting, whether to display animated emoticons or not, statistics collected, etc. All this information is stored in clear-text.

Skype also stores the user's contact list at a central server so that if the user logs in from a different machine, the contact list is still available. In earlier versions of Skype, the contact list was stored locally only, limiting a user to login from one machine. If a user logged in from another machine, the only way s/he could connect to one of his/her contacts was by remembering the user name for the contact and re-entering that information in the local client.

9.3.2.2 *Login and authentication*   The Skype login processes the SC connecting to a SN and then verifies its identity to a centralized authentication server that stores each user's password. The HC of the nodes stores the IP addresses and port numbers of SNs that have been recently active. The SC tries to connect to one SN at a time until it finds one which is active. The SC must be able to establish a TCP connection with the SN. If none of the SNs in the HC is reachable, then the SC tries to connect to some SNs whose address is hard-coded in the Skype executable. These SNs also form part of the Skype infrastructure.

Once SC connects to SN using a TCP connection, all its information exchange with the Skype network is through the SN. In fact, it has been observed that the SC routes the login messages through the SN. For more details the reader is referred to [1]. This has the effect of reducing the probability of Skype being disabled through a simple firewall which restricts packets going to any of the Skype login servers. This is achieved as the identity of SNs keeps on changing, thus reducing the likelihood that a Skype message to all potential SNs would be blocked.

9.3.2.3 *Global index*   After a user is authenticated by the login server, the SC informs the Skype network of its presence through its SN. The information regarding the availability and the location of the user is stored in the overlay network of SNs. This structure is called Global Index. Essentially, Global Index is a distributed directory of users in the Skype network. Maintaining this directory in a centralized fashion is unscalable and requires a significant amount of infrastructure to support a large user-base.

The Skype design of Global Index utilizes the resources provided by the end-hosts to share the burden of directory maintenance among the network participants. This insight made it feasible to support a huge user-base using Skype. The specific details of the Skype Global Index have not been made public. However, it is known that the Global Index is a multitiered organization structure where SNs can communicate with each other. The objective is to reduce the latency with which any SN can locate any user. The claim is that Global Index ensures that the information regarding any user who logged into the network during the previous 72 hours is always available, irrespective of the joining and leaving of SNs during that period.

9.3.2.4 Call setup and routing
In order to set up a call, a user clicks on the contact name with whom s/he wants to communicate. The SC looks up the status of the contacted ID in the overlay network through its associated SN. The SN uses the Global Index to report the current IP address and status of the contact. The SC then initiates a direct call to that user's IP address and port. The call setup also involves exchange of identity information among SCs that both the SCs validate. This ensures that no SC is impersonating some other entity.

While the direct connection is a simple solution for VoIP calls, it may not result in the best quality in all scenarios. Skype realizes this and performs specialized routing over the P2P overlay in order to reduce the latency and minimize the packet losses. As discussed in the previous chapters, an overlay can be used to enhance the VoIP quality using path diversity and coding. Skype utilizes these resources well to enhance the VoIP quality. It even uses the path-switching method where it keeps multiple connection paths open and, depending on the prevalent network conditions, it chooses the best current path for the call. This improves call quality with reduction in latency.

9.3.2.5 NAT traversal
Not all host machines connected to the Skype network have a public IP address. These machines are behind NAT devices which themselves have publicly routable IP addresses but assign local IP addresses to the host machines. The NAT devices relay traffic between the host machine and the Internet by acting as an intermediary. In order for machines behind NATs to be able to use Skype, calls made to these machines must be routed behind the NAT to the concerned machine.

Skype uses a proprietary method to detect whether the node is behind a NAT and if so, what type of NAT it is behind. It then uses publicly routable nodes as NAT traversal aides to determine the type of the NAT and accordingly relays traffic between the Skype network and the host behind the NAT device. As shown in Figure 9.3, if clients SC_1 and SC_2 are behind NATs or firewalls that prevent incoming connections routed to these nodes, then a relay SN is used to act as a rendezvous point for the traffic between these nodes. The prevalent thought is that the Skype NAT traversal logic is similar to the standardized STUN and TURN protocols.

The choice of node that can be used as rendezvous points for such nodes that are behind the NAT requires that the nodes be publicly routable and that they have sufficient resources. A subset of nodes from among the SNs serve as the rendezvous points. The techniques designed do not have a significant overhead and have allowed Skype to break through several types of NAT device.

9.3.2.6 Conferencing
While the previous sections deal with the communication between two parties, Skype allows multiple users to communicate as part of a single conference. Skype does not support a full-mesh-based conference. In a conference of n

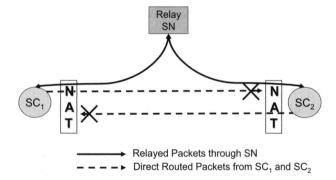

    Relayed Packets through SN
- - - - ▶ Direct Routed Packets from SC_1 and SC_2

Figure 9.3   The NAT devices in front of both SC_1 and SC_2 make it impossible for them to route packets to each other directly. They can establish communication with a publicly routable SN that can relay packets between them.

participants, not all nodes need to be directly linked to each other using overlay links. Instead, the data received at a high-resource node is relayed out to other nodes. Thus, one node can receive stream from two other nodes and then along with playing it back to its user, it can relay the streams to the participant nodes. This sort of a structure can be thought of as a hub-based mode.

The conferencing procedure between three nodes SC_1, SC_2, SC_3 is shown in Figure 9.4. Of the three nodes, one with the maximum resources (say SC_1) becomes a relay node. SC_2 and SC_3 do not need to forward the packets to each other directly.

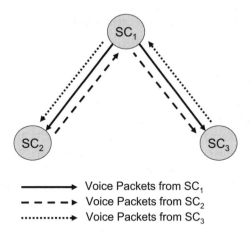

    Voice Packets from SC_1
- - - - ▶ Voice Packets from SC_2
·············▶ Voice Packets from SC_3

Figure 9.4   Conferencing among three SCs in Skype. SC_1 acts as the relay node to forward the voice packets from SC_2 and SC_3 to the other.

9.3.3 Encryption

Maintaining privacy of information is a first-order design goal in the Skype network. Thus almost all messages exchanged in the network are encrypted end-to-end. Furthermore, Skype uses standard-based implementations of strong ciphers to keep the information secure. We list some of the techniques used in Skype in this section. For more details on the use of encryption in Skype, the reader is referred to [2].

The basics tools used in Skype are AES (Rijndael) block cipher, RSA public-key cryptography, ISO9796-2 signature padding scheme, RC4 stream cipher and SHA-1 hash functions.

The core secret of the Skype system is the private signing key of the Skype's server. The public key for this signing key is stored in the Skype binary. When a user registers with the Skype network, the client generates a RSA key-pair. The private portion of the key along with a hash of the password are securely stored on the user's machine. Then an AES encrypted session is established with the server. Over this connection the new username and the hash of the password are sent to the server which verifies the uniqueness of the username and stores a hash of the password that the user sent. The server creates and signs an identity certificate for the user. The certificate binds the user name to the corresponding key and is signed by the server's certificate. Depending on whether the user has subscribed to the premium services or not, the server signs the certificate with a longer 2048-bit key or a simpler 1536-bit key.

The connection between two end-hosts also involves exchange of the user identity certificates over a secure connection. To establish this connection the clients challenge each other with random nonces. The other party signs the challenge with his/her own private keys. These can be verified using the public key stored in the identity certificates that were exchanged. The shared key for communication is established by each client providing 128 bits for a collective 256-bit AES key. Using this AES, a stream of keys is generated that is XOR-ed with the basic data to produce encrypted data.

An independent study done on the invitation of Skype for its security evaluation raves about Skype's cryptography standards [2]. The strong emphasis on security in Skype ensures that the information transferred and authenticated is genuinely safe to a high degree.

9.3.4 Skype performance

Since there is very little information in the literature about the fast-track protocol, the studies focusing on the behavior of the Skype network are quite limited. Also, since all communication in the network is encrypted, it is not possible to look at the content of the packets and figure out the behavior of the users and the super-nodes in the Skype network.

Due to these constraints, the studies of Skype behavior have looked at the macro-level properties of the network. Despite the limited information that is available, these studies have come up with significant insights into the behavior of Skype protocol. In this section, we present the results of some of these studies [2, 4].

The experiments have been carried out on real Skype clients and several important issues were identified and addressed. The goals of the study were to capture the behavior of the Skype network over time and the quality of the Slype client under different traffic conditions.

To establish the justification of using a SN-based architecture as the basis for storing the distributed Global Index, it is important to ascertain that the network of SNs is relatively stable. If the network was highly volatile, it is unlikely that the contents of the Global Index may be distributed and searched efficiently. To this end, the network activity of a SN was recorded for several days. It was observed that there are more than 250 K SNs in the world with portions of the network remaining active at different times.

An experiment to measure the stability of the Skype SNs was also carried out during the course of the study. Of the large number of Skype SNs, 6000 were randomly chosen to monitor. Every few minutes, simple application level pings were sent to these machines and the information was recorded. It was observed that the churn in Skype network is quite low. This is in contrast to the user-base in the same P2P network for file downloads where the users keep logging on and off. In the VoIP setting, the users tend to let the software run for longer durations. The mean session duration is of the order of several hours. The number of servers online shows a diurnal trend where more users are online in the morning and fewer in the evening. This trend is also visible at a coarser granularity of timescale where there are more servers online during weekdays than during weekends.

The first reaction to using SNs to relay traffic for host behind NAT would be that the overheads of such a method would be prohibitive. It has been observed that for more than 50% of the time, one supernode uses a bandwidth of less than 205 bps.

It has been reported in multiple forums that Skype does not use silence suppression, resulting in approximately 30 packets per second being sent into the network irrespective of whether there is a talk spurt or silence on the line. Clearly, using silence suppression, simple technique could reduce the client bandwidth consumption. This would have a further impact in reducing the overhead on the supernodes, which may end up relaying traffic between pairs of client nodes that are not directly accessible from each other.

Another observation that has been made is that the average length of a VoIP call is around 13 min. This average duration is much higher than that for the traditional telephony network where the mean duration is only 3 min. Perhaps this can be attributed to the high quality of the calls and the PC-to-PC calls being free. This can be thought of as fundamentally shifting the traffic profile merely by the features provided by a single application.

Another set of experiments on the performance of the Skype client are reported in [4]. Skype's performance is tested in a controlled environment where the network conditions are varied using the NIST Net network emulator. The quality of voice delivered under various network circumstances is measured using the PESQ, metric which compares the original and received audio to give a quality score.

The Skype client achieved a reasonable PESQ MOS score of 3.5 when the network traffic capacity was 50 Kbps. The score goes below 3 with rates of 40 Kbps or less, indicating that the network should have at least this amount of bandwidth for Skype to deliver decent quality VoIP. The PESQ MOS values degrade almost linearly with the available capacity.

When the network has sufficient capacity of 50 Kbps and the network delay increases, the conversational quality decreases. This happens because the mouth-to-ear latency becomes perceptible to humans beyond a certain delay. In the experiments the Skype client tended to reduce the transmission rate when the latency went beyond the acceptable limit.

When the network losses are increased, the Skype client tries to adapt by increasing the transmission rate used so as to compensate for the losses. It may use specialized encoding as we described in the context of VoIP over infrastructure overlays. The net result is that the PESQ MOS remains reasonable and can withstand up to a 10% loss in the experimental setting.

In summary, the experiments and analysis suggest that the Skype client tries to adapt to capacity, latency and loss to improve the VoIP quality in the data plane. The Skype SN network is quite large and stable, so that the maintenance of Global Index can be carried out reasonably well. This gives credence to the claim that if any user has logged into the

Skype network in the past 72 hours, the information about that user is not lost in the Global Index.

9.4 STANDARDIZATION

The idea of having VoIP in a P2P overlay has gained considerable attention over the last few years. Recently, there has been a concerted effort in standardizing the requirements for P2P-based VoIP that uses SIP as the base control plane protocol. This effort has resulted in the formation of the P2PSIP working group in IETF.

The standardization of SIP in a P2P network is in its infancy at present. To date, there are no standards or proposed standards in this space. Several preliminary Internet drafts have been written with the objective being to put the possible standardization candidates in the open for people to debate. The goal is to define a set of protocols that enable SIP to use P2P methods to perform several tasks, including resolution of the target of SIP requests and message transport.

At a high-level, P2PSIP intends to use an overlay network of nodes that collectively behaves as a distributed database. This database stores information about resources in the form of *Resource Records*. It allows insertion, lookup and deletion of Resource Records based on a unique Resource Record identifier. Resource is a broad term which can include several entities such as users and services. The Resource Record for each resource contains resource-specific information and the Resource Record Identifier is derived from the resource potentially as a unique hash of some subset of the identifying values stored in its Record. As of July 2007, no decision has been taken as to how this distributed database functionality would be implemented. The general direction is to use a DHT that provides efficient insertion, lookup and deletion of records.

The exact use method of this distributed database has not yet been standardized. However, it is likely to follow a simple logic where a node (peer X) registers the location of a user U by inserting in the database its own address as the contact point for U. When another user with contact point as peer Y wants to contact user U, then peer Y looks up the DHT for the current location of U. In response it obtains the address of peer X. Then, a standard SIP-INVITE method is issued to user U and location X. Then, the communication continues as in standard SIP protocol. There are variants of this procedure that have been proposed but none has yet been standardized.

The P2PSIP working group is also addressing the other architectural issues such as NAT traversal, client and peer behavior, interaction with non-P2PSIP entities, and security considerations. For details, the reader is referred to the current Internet draft on P2PSIP [5].

9.5 SUMMARY

Several issues arise when we deploy VoIP over a P2P overlay. These issues can be broadly classified into those that are due to the architectural considerations and those that arise because of the P2P network properties. Ensuring privacy and availability of information are the two most important criteria that will determine whether or not a P2P VoIP service will be usable. Skype solves these issues by using end-to-end encryption in almost all communications and making the information highly available by using proprietary Global Index technology. Experimental results show that the Skype P2P network contains nodes that remain in the network for long durations indicating that the network is quite stable for the Global Index to work well. The Skype client tries to adapt the quality of the voice when faced with adverse network conditions. The opportunity observable in deploying VoIP in a P2P setting has already given rise to standardization activity in this space in the form of the P2PSIP working group of the IETF.

REFERENCES

1. Baset, S. and Schulzrinne, H. An Analysis of the Skype peer-to-peer internet telephony protocol, *Proceedings of IEEE Infocom* (2006).

2. Berson, T. Skype security evaluation, *http://www.skype.com/security/files/2005-031 security evaluation.pdf* (2005).

3. Guha, S., Daswani, N. and Jain, R. An experimental study of the Skype peer-to-peer VoIP system, *Proceedings of IPTPS* (2006).

4. Barbosa, R., Kamienski, C., Mariz, D., Callado, A., Fernandes, S. and Sadok, D. Performance evaluation of P2P VoIP application, *Proceedings of ACM NOSSDAV* (2007).

5. Bryan, D., Matthews, P., Shim, E. and Willis, D. P2PSIP concepts and terminology, *http://www.p2psip.org/drafts/draft-ietf-p2psip-concepts-00.txt* (2007).

PART III

VOIP IN WIRELESS NETWORKS

Today, there exist several choices of wireless network that provide the users (mobile and fixed) access to Internet or corporate Intranet. A major advantage of wireless access networks is in supporting mobility to VoIP users and the ease in terms of providing network coverage. However, wireless access networks differ in terms of their capabilities in supporting VoIP. For example, existing 3G cellular networks cannot support VoIP using their packet-switched connections albeit providing voice to users through a traditional circuit-switched approach. Another example is the 802.11-based WiFi network, which, even though it provides a high data rate, is not well equipped to provide the QoS guarantees required by VoIP.

Clearly, it is worthwhile to understand the features and the limitations of the wireless access networks. An understanding guides the design of network or application level solutions that are required to enable the delivery of VoIP over these networks.

The chapters in this section provide the reader with a thorough understanding of two important wireless access technologies that are well positioned today to carry VoIP service to users: IEEE 802.11-based Wireless LAN (WLAN) and IEEE 802.16-based WiMAX. While WLAN is typically being deployed to provide short-range coverage such as in the office and the home, WiMAX is being deployed to provide wide-area broadband wireless access to mobile users.

VoIP: Wireless, P2P and New Enterprise Voice over IP   Samrat Ganguly and Sudeept Bhatnagar
© 2008 John Wiley & Sons, Ltd

The focus of the chapters on these two wireless access technologies is motivated by the following reasons: (a) both of these technologies expose a uniquely different level of features and characteristics; and (b) both technologies are used as a packet-switched IP network, thereby making them a natural choice for supporting VoIP. The chapters first provide a thorough understanding of the technology and subsequently provide various techniques and solutions being proposed specifically to support VoIP.

CHAPTER 10

IEEE 802.11 WIRELESS NETWORKS

In the recent past, there has been a tremendous increase in the deployment of IEEE 802.11-based wireless LAN (WLAN). Although initially a WLAN was planned to serve as a hot-spot for high-speed data coverage, WLAN technology is becoming increasingly popular for supporting VoIP both at home as well as in an enterprise environment.

The main advantages driving the adoption of Voice over WLAN are easy, non-intrusive and inexpensive deployment, low maintenance cost, universal coverage and basic roaming capabilities. In this chapter, we provide a brief primer on the IEEE 802.11 wireless network technologies. The goal is to establish a solid background for the next chapter where the focus is on understanding the performance related problems and solutions when deploying VoIP over WLAN.

10.1 NETWORK ARCHITECTURE OVERVIEW

We start with a description of the IEEE 802.11-based WLAN by describing the components involved in the network architecture. These components can interact in different ways depending on how the WLAN is configured. We will give an overview of these possible configurations and of how components interact in these settings.

10.1.1 Components

Any device such as a laptop or a handheld that is equipped with a IEEE 802.11-based wireless Network Interface Card (NIC) is referred to as a *station*. A Station is an IEEE

VoIP: Wireless, P2P and New Enterprise Voice over IP   Samrat Ganguly and Sudeept Bhatnagar
© 2008 John Wiley & Sons, Ltd

802.11 network client that communicates with the IEEE 802.11 wireless network. In most common deployment scenarios, a station connects to an *Access Point (AP)* and gains access to the external network beyond the AP. Typically, access points are devices that serve multiple stations directly and act as a bridge between the IEEE 802.11-based wireless network and the wired counterpart (typically the Internet).

The physical region that a single access point can cover is limited essentially to a circle centered with itself and a diameter of about 200 m. The region in which an IEEE 802.11 node can communicate with another node is called its *Range*. The limited range of an AP implies that covering a larger region such as an entire floor in a building requires deployment of multiple access points. A network of multiple access points is called a *Distribution System*. A Distribution System has to keep track of stations for each AP and has to forward data frames correctly between the stations and the gateway to the external network.

10.1.2   Network configurations

There are three basic types of network configuration for IEEE 802.11 wireless networks. Network configurations are based on the communication pattern between the stations and the external wired network.

10.1.2.1   Ad hoc networks   The ad hoc network configuration in an IEEE 802.11 network facilitates a communication pattern in which the stations communicate directly with each other without the involvement of an AP. Figure 10.1(a) shows the ad hoc network configuration. Ad hoc networks typically share the following properties. Each station can communicate directly to only those stations that are in its range. Ad hoc networks are typically created on-demand and are mostly dynamic depending on the user participation. The network keeps on changing as the users who participate in it keep joining and leaving. IEEE 802.11 ad hoc networks are identified by a name that is referred to as Independent Basic Service Set (IBSS).

10.1.2.2   Infrastructure networks   The most popular network configuration in terms of the deployment architecture is the infrastructure network mode. Infrastructure networks always employ access points to connect the stations to the IEEE 802.11 wireless networks. In other words, a station cannot communicate directly with another station or the external network without going through the access point. Thus, in order to gain access to the IEEE 802.11 network, a station must be in the range of an access point and has to associate with an access point. Figure 10.1(b) shows a typical infrastructure network configuration.

There can be a group of access points covering a large area while belonging to a single administrative domain. Such a network of access points is called an Extended Service Set (ESS). An access point belonging to a given ESS is identified by a Service Set Identifier (SSID). Stations trying to access a given network associate only with the APs having the corresponding SSID. Furthermore, stations can only talk to the other stations that belong to the same ESS and share the same SSID.

10.1.2.3   Infrastructure mesh networks   Typically, a distribution system connecting the access point for a given ESS is connected using wired backbone. However, in many coverage scenarios where wired backbone is either difficult to deploy or expensive, a multi-hop wireless backbone is advantageous. Such a multi-hop wireless backbone is

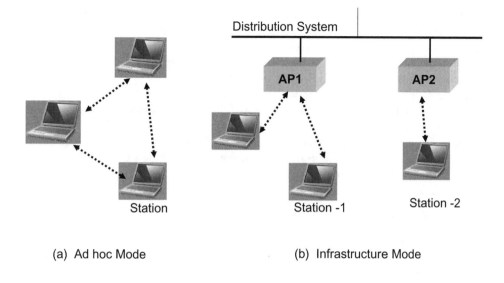

Distribution System

AP1   AP2

Station   Station -1   Station -2

(a) Ad hoc Mode   (b) Infrastructure Mode

Figure 10.1   Basic network configurations.

called a wireless mesh network. Using a wireless mesh network to connect the access points is commonly referred to as the Infrastructure Mesh Network or simply an IEEE 802.11 mesh network. In such a network, a station connects to an access point and is unaware of the wireless mesh backbone. The mesh backbone is responsible for forwarding the data intended for a given station to the correct access point.

10.2 NETWORK ACCESS MANAGEMENT

The network access management functionality ensures that stations can enjoy seamless access to the wired network through the access points. In order to provide such seamless access, IEEE 802.11 infrastructure wireless networks provide a set of services and define a set of protocols. We describe these services and protocols next.

10.2.1 Association

The first important operation for a station is to become associated to a given access point. Association requires that the station be in the range of the AP. Association starts with the scanning or the discovery phase where a station passively scans the beacons transmitted by the APs.

The beacons from an access point are sent periodically (typically once every 100 ms) and are primarily used to establish and maintain connection between the associated stations and the AP. A single beacon frame contains time stamp information to update the local clock at the station. Other beacon information includes the SSID value to identify the network and the supported data rates (11 Mbps, 5.0 Mbps) along with other parameters.

A station passively scans the beacons emanating from different APs on a given radio frequency (RF) channel. By listening to the beacons, the station can decide which AP to join. The process of joining or becoming connected to an AP is known as *association*. As a part of the association process, the station chooses which AP to join. The choice is based on the SSID that identifies the network and the BSS that identifies the particular AP for a given ESS. Choice of AP may be based on the signal strength as reported by the beacon received by the station. Typically, connecting to the AP with the highest signal strength is recommended for most application scenarios. However, for other purposes such as load balancing, a different selection criteria might be used.

Once the choice of an AP is made, the station initiates the actual association process by issuing an association request frame. If the association is granted, the AP returns to the station with a status code (0 = successful) and Association ID (AID). AID is required to identify the station and is used for forwarding the frames to the station. Once the association is complete, the AP starts transferring data frames to/from the station.

10.2.2 Authentication

In a typical IEEE 802.11 wireless network, the association process is initiated only after a successful authentication process. The authentication process is initiated by the AP by issuing an authentication request. In the authentication process, the identity of the station is validated and the station is authenticated for access.

In the simplest form of authentication process (referred to as *open*), the station provides the MAC address to the AP. A network administrator can provide the MAC address-based filtering rules to restrict the access to certain stations. Note that the MAC address can be easily cloned and is a very weak form of providing access control.

A more widely used authentication mechanism is the shared WEP key-based authentication. In this mechanism both the AP and the station share a common key. During authentication, the AP sends a challenge text to the station. The station encrypts the text using the Wired Equivalent Privacy (WEP) key and returns it to the AP. If the AP can decrypt the text using its own key, it proves that the station has the same key. Therefore, successful authentication amounts to proving that the station has the right shared WEP key. A static WEP key can be easily obtained by an entity that can hear the packets. A stronger authentication process utilizes a Radius server based on 802.1x protocol where the WEP key is dynamically generated.

Recently, security researchers have come up with methods to break the WEP key regardless of whether it is static or dynamic. It is possible to retrieve the key using a few thousand packets. Thus, the recommended practice is to use stronger mechanisms such as *Wi-Fi Protected Access* (WPA and WPA2) that implement the IEEE 802.11i standard. The main benefits of these schemes are that they keep changing keys periodically. WPA2 has a stronger encryption using an Advanced Encryption Standard (AES) based algorithm that is widely believed to be very secure.

10.2.3 Mobility

When a user moves from one location to another, s/he can leave the coverage range of the current (old) AP and move into the coverage range of a new AP. The above action results in a *connection handoff* from the old AP to the new AP. To initiate hand-off, the station is required to perform a number of steps, which are discussed next.

The basic IEEE 802.11 wireless network standards do not define any specific hand-off mechanism. Consequently, the hand-off process is implementation-specific. In a typical scenario, hand-off is initiated by the station by periodically monitoring the signal strength.

If the signal strength of the packets received from the current AP is weak, the station starts scanning for new APs. If the station finds a new AP with better signal strength, the station sends a reassociation request to the new AP. Reassociation requests carry information about the old AP. The new AP responds with a new association ID. The new AP also contacts the old AP and gets the buffered (undelivered) frames from the old AP to be subsequently delivered to the station.

It must be understood that such hand-off works seamlessly when the new AP belongs to the same ESS and shares the same SSID with the old AP. Typically, a particular ESS is built on the same VLAN, thereby requiring no extra modifications in packet-forwarding inside the distribution system. If, however, a user moves across ESS where two networks have different IP addresses, then mobile IP-based solutions are required to facilitate a seamless hand-off.

10.3 BASIC MEDIUM ACCESS PROTOCOL

The IEEE 802.11 *Medium Access Control (MAC)* protocol defines the method for accessing the shared wireless channel for transmitting data. The main goal of the MAC is to ensure that multiple stations can transmit data on the same wireless channel without causing any interference. This section provides a brief overview of the MAC layer functionalities. The standard IEEE 802.11 MAC protocol that is referred to as Distributed Coordination Function (DCF), is based on Carrier Sense Multiple Access with Collision Avoidance (CSMA/CA).

10.3.1 Distributed Coordination Function (DCF)

As the name suggests, the DCF protocol is a distributed protocol requiring no centralized entity for coordinating the channel access. In the DCF protocol, there is no differentiation between APs and station in terms of functionalities. Therefore, we use station to denote just a node participating in the DCF protocol, which includes both stations and APs.

There are two parts in the design of the DCF protocol. The first part defines the mechanisms used for sensing the channel. Through sensing the channel, a station determines if the channel is busy or is used by another station for data transmission. When a channel is sensed as being busy, all stations attempting to transmit data defer the channel access. The second part of the DCF protocol provides a basic mechanism that prevents more than one station from deciding to transmit data at the same time. Both mechanisms assist in assuring interference or collision-free data transmission with multiple contending stations.

10.3.1.1 Carrier sensing There are two approaches for carrier sensing that identify whether the channel is busy or idle. These approaches are referred to as (a) *physical carrier sensing* and (b) *virtual carrier sensing*. In physical carrier sensing, the physical layer monitors the received signal strength, which is also referred to as the *Clear Channel Assessment* (CCA) function. If the observed signal strength is above the pre-defined carrier sense (CCA) threshold, the medium is considered to be *busy*, otherwise *idle*.

In the case of virtual carrier sensing, DCF protocol uses the Network Allocation Vector (NAV) for the station that acquires the channel for data transmission. The NAV provides the amount of time required by the transmitting station to complete its current data transmission (in terms of the number of slots). The transmitting station includes the NAV value in the duration field of the header of the frame. When any station receives the NAV value, it starts a timer with the NAV value and subsequently defers transmission until the timer expires. Using the NAV value, the length of transmission is advertised to all contending stations that are attempting to transmit.

10.3.1.2 *Random access*

The carrier sensing mechanism ensures that a station does not start transmitting when the channel is busy. However, when the channel becomes idle, more than one proximate station can sense the channel's idle status at the same time and start transmitting data. This would result in packet collision. In order to prevent such a scenario, the DCF protocol employs a random back-off mechanism, as was originally proposed in the CSMA protocol. According to this mechanism, a station attempting to transmit data chooses a random slot (k) in a window (W). The station transmits only if the channel is idle from the first until the kth slot. In other words, among a set of contending stations, the station with the lowest value of random slot gets the chance to transmit in the given window.

Unfortunately, if the window is too small, there is a high probability that more than one station chooses the same random slot. In order to deal with this situation, the window (W) is doubled each time there is a packet collision. Clearly, increasing the window size reduces the probability of collision at the expense of lower throughput and increased delay. Therefore, the window is reset to the minimum value once a successful packet transmission is achieved.

One should also note that such a random access scheme cannot ensure strict QoS guarantees and can result in increased jitter for VoIP traffic. However, the random access mechanism in DCF protocol ensures fairness in channel access in the presence of low to medium traffic load.

10.3.2 Station protocol

When a station has no data to send, it is in the idle state and checks if there is any data arriving from the higher layer. With data to transmit and the channel idle, the station waits for a specified duration called DCF Inter Frame Space of 50 μs (DIFS). If the channel becomes busy during DIFS, the station waits until the channel becomes idle and waits again for DIFS amount of time. On the other hand, if the channel remains idle for the DIFS amount of time, the station enters the contention or BACK-OFF period where the station waits for a random number of slots from a window W. Therefore, the station starts a BACK-OFF timer with the onset of the contention period. The BACK-OFF timer is frozen if the channel becomes busy and resumes when the channel becomes idle for a DIFS amount of time. At the expiry of the BACK-OFF timer, the station transmits packets.

The DIFS amount of waiting time results in a delay. Therefore, a smaller duration Short Inter Frame Space of 10 μs (SIFS) is also used in conjunction with DIFS for certain types of priority packets such as ACKS or VoIP packets. Figure 10.2 illustrates a simple case of IEEE 802.11 data transmission protocol between sender and receiver.

Once a station gets hold of the channel, it transmits the entire packet. When the destination station receives the packet, it waits for SIFS amount of time before sending an ACK. Receiving the ACK, the sender knows that the packet has successfully reached the destination. If the ACK is not received, the sender retransmits the packet. The basic version of the station protocol is shown in Figure 10.2.

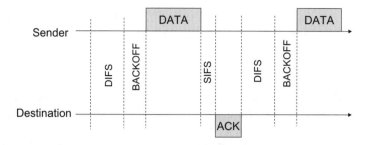

Figure 10.2 A simple case of IEEE 802.11 basic medium access protocol.

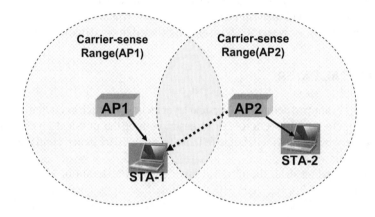

Figure 10.3 Example of hidden terminal problem.

10.3.3 Hidden terminal problem

The carrier-sensing mechanism only works when two transmitters can hear (carrier sense) each other. In a scenario where there are two transmitters who cannot detect each other's presence through carrier sense, they will start transmitting data simultaneously. Such a scenario results in packets from both transmitters reaching a receiver at the same time and causing packet collision. This problem is called the *Hidden Terminal Problem*. The above scenario is illustrated in Figure 10.3. Typically, the current sophisticated receivers can decode a packet from the correct transmitter as long as the packet from the interfering transmitter is received with a significantly lower signal strength compared to the desired packet. This phenomenon where the receiver is able to sift through the interfering packet to decode another packet is referred to as the *capture effect*. It should be observed that the effect of multiple interferers can be characterized by adding the signal strengths from the interferers at the receiver.

In order to prevent the hidden terminal problem, DCF protocol defines the use of *Request-to-send* (RTS) and *Clear-to-send* (CTS) frames in the following way. Before transmitting data, the station sends an RTS frame advertising its intent to transmit data. After receiving an RTS, the receiver station sends a CTS frame acknowledging that it is ready to receive the data and allowing the sender to start transmission. By hearing the CTS frame, all stations in the vicinity of the receiver are silenced, thereby preventing the hidden terminal problem. However, it is to be noted that use of RTS/CTS incurs high overheads when used with small packets such as VoIP. Typically, RTS/CTS is not enabled in deployed IEEE 802.11 networks owing to lower throughput from the high overhead.

10.3.4   PCF

Another optional MAC approach defined under the IEEE 802.11 protocol is called the *Point Coordination Function* (PCF). PCF is a centralized protocol and requires the existence of a centralized coordinator entity called a point coordinator. Point coordinators reside inside an access point and decides which station can transmit in a given time period. Based on this decision, the point coordinator polls each station. When polled, the given station can transmit data in a contention-free environment. The salient feature of the PCF is that it is easy to provision QoS with deterministic delay bounds, which is advantageous for voice and video applications. However, PCF is seldom enabled in the deployed networks.

10.4   PHYSICAL LAYER

The IEEE 802.11 standard defines the physical layer techniques such as coding, modulation, carrier sensing, etc. In this section, a brief description of the physical layer behavior of IEEE 802.11 network is provided. Some of these techniques are generic and are applicable in any wireless network.

In order to transmit data, the physical layer encodes a stream of bits into a stream of symbols. Each symbol may consist of multiple bits. Subsequently, the symbols are converted into analog RF waveforms in the modulation process for transmitting the signals. The physical layer imposes a frame structure on the symbol transmission to enable the receiver to become aware of when to demodulate the received signal. The start of a frame is identified by sending a preamble, which is a predetermined sequence of symbols transmitted at the beginning of each frame.

The receiver demodulates the received signal and performs a symbol level decoding to reconstruct the data transmitted by the sender. For the receiver to be able to decode a signal, the signal-to-noise ratio (SNR) should be high enough. Loss in signal strength with distance or the path loss along with multipath fading are the main causes for lowering the SNR at the receiver. For a given SNR, the decoding capability of the receiver is dependent on the channel coding and modulation technique used. A brief overview of the coding and modulation techniques used in IEEE 802.11 is provided next.

10.4.1   Spread spectrum techniques in IEEE 802.11b

The IEEE 802.11b physical layer uses spread spectrum-based techniques for encoding the symbols. In this technique, a narrow band data signal is spread over a wider band before transmission. Spreading is performed by multiplying a spreading sequence (a binary code) with the input stream. IEEE 802.11b implementation uses an 11 bit spreading code. This form of spreading is referred as *direct sequence spread spectrum* (DSSS). Since the spreading sequence is generated at a higher frequency, the above multiplication results in increasing the original signal rate spanning a greater amount of frequency bandwidth.

It must be noted that multiple signals can be transmitted on the same frequency band by choosing different spreading sequences. The receiver can correctly decipher the correct signal by knowing the corresponding spreading sequence. The coexistence of multiple signals ensures that the spectral efficiency remains high. At the same time, spreading a single input signal over a wider band ensures that the signal is less susceptible to noisy channel conditions.

10.4.2   OFDM in IEEE 802.11a

IEEE 802.11a is based on *Orthogonal Frequency Division Multiple Access (OFDM)* technology. OFDM can result in higher throughput and spectral efficiency. Furthermore, OFDM can minimize multipath propagation problems. In OFDM, the frequency band is divided into smaller size bands called subcarriers. The subcarriers in OFDM can be overlapping. Typically, a particular transmission to a given station uses a subset of the subcarriers. In the case of IEEE 802.11a operating at 5.2 Ghz, a 20 MHz channel is divided into 48 subcarriers providing a maximum data rate of up to 54 Mbps. Basically, while in DSSS, a single symbol is sent at a given time over the channel, in OFDM, 48 symbols are sent in parallel on the 48 subcarriers. IEEE 802.11g uses a combination of DSSS and OFDM operating at 2.4 Ghz. This combination also provides a data rate of up to 54 Mbps.

10.4.3   MIMO in IEEE 802.11n

In order to provide an even higher bandwidth, the recently proposed *Multiple Input Multiple Output (MIMO)* technology uses a concept called *antenna diversity*. MIMO techniques leverage spatial diversity by having multiple transmitting and receiving antennas per station. Through multiple antennas, multiple paths are created between a transmitter and a receiver, thereby providing spatial diversity. The input signal is then coded over multiple paths to either increase spectral efficiency (to increase throughput) or robustness (to increase reliability against noise). The MIMO technique used in IEEE 802.11n is targeted to provide peak raw data rate beyond 100 Mbps.

10.4.4   Modulation and rate control

Modulation converts the symbols into analog waveform. The coding of the symbols is performed either by phase-shifting values or by amplitude levels. Depending on the number of levels used, more symbols or information can be packed, resulting in a higher bit-rate. For example, with differential binary phase shift keying (DBPSK) using two phase shifts can achieve 1 Mbps bit-rate in IEEE 802.11b. On the other hand, differential quadrature phase shift keying (DQPSK) with four possible phase shifts attains a 2 Mbps bit-rate. Table 10.1 shows bit-rates for some prominent modulation schemes.

For a given SNR, high bit-rate modulation results in higher bit-error probability or packet loss. Figure 10.4 shows the bit-error probability with different SNR for different modulation rates. Therefore, the effective throughput observed at the receiver is a function of the bit-rate used and the SNR. For example, if the SNR is low, it is better to use a lower bit-rate. However, for a higher SNR, a higher bit-rate provides higher effective throughput. In order to increase the effective throughput, modern IEEE 802.11 devices uses auto-rate control where a correct bit-rate is chosen automatically by sensing the channel conditions. Such bit-rate adaptation works by measuring the delivery ratio during a period of time and adjusting the rate accordingly.

Table 10.1 Bit-rates supported by different variants of IEEE 802.11 along with the coding and modulation used.

Bit-rate in Mbps	IEEE 802.11	Channel coding	Modulation
1	b	DSSS	BPSK
2	b	DSSS	QPSK
5.5	b	DSSS	CCK
11	b	DSSS	CCK
6	a/g	OFDM	BPSK
9	a/g	OFDM	BPSK
12	a/g	OFDM	QPSK
18	a/g	OFDM	QPSK
24	a/g	OFDM	QAM-16
36	a/g	OFDM	QAM-16
48	a/g	OFDM	QAM-64
54	a/g	OFDM	QAM-64

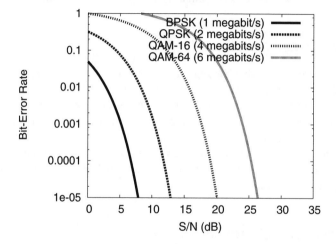

Figure 10.4 Example of hidden terminal problem.

10.5 NETWORK RESOURCE MANAGEMENT

Resource management is a critical part of an IEEE 802.11 wireless network due to the fact that the wireless medium is shared among multiple access points and stations. Resource management tries to address various performance-related concerns and is an area of active research. To gain a better understanding of the issues that need to be addressed, we first give a general overview of the interference model for IEEE 802.11 networks. Subsequently, the mechanisms for resource management are detailed.

10.5.1 Interference model

An interference model characterizes the capacity of the wireless link between a sender (AP) and a receiver (Client) or vice versa. A wireless link is formed between an AP and a client only if the client is in the transmission range of the AP. Although the raw capacity is determined by the specific IEEE 802.11-specific physical layer technology used, the actual capacity is more interference dependent. Interference is caused by the active interfering transmitters (APs or the clients) in the neighborhood of the sender or the receiver. For an easy exposition of the interference model, consider two types of interference – *sender-side interference* and *receiver-side interference.*

Sender-side interference results in a decrease in the sending throughput of a sender in the presence of active transmitters in the neighborhood of the sender. Sender-side interference only happens when the sender is in carrier-sense range of the interferer and can sense the channel as being busy when the interferer transmits. In this case both the sender and interferer contend for access of the medium in order to transmit packets. However, the IEEE 802.11 DCF employing random access mechanism ensures that each contending transmitter has an equal opportunity to use the medium. Therefore, the data sending rate is almost equally divided among the contending transmitters. For example, a single IEEE 802.11b AP has a maximum sending rate of 5 Mbps when there is no contention. However, if the AP is contending with another AP, the rate reduces to approximately 2.5 Mbps. In other words, the medium is shared equally among the APs.

Receiver-side interference reduces the wireless link throughput by increasing packet collisions. Receiver-side interference is a result of the hidden-terminal effect when the interferer is outside the carrier-sense region of the sender. This implies that the interferer does not back off when the sender transmits. Packets from sender and interferer reach simultaneously at the receiver, resulting in packet collision at the receiver and lowering the effective link throughput. Capture effect at the receiver is likely to alleviate the impact of such interference in some instances but it remains a major concern in a wireless network.

The interference effect is more pronounced in a dense AP deployment. On one hand, an increase in AP density increases the maximum throughput to a client by reducing the expected distance between an AP and client (increasing the expected received signal strength). On the other hand, increased density can reduce the sending rate and packet reception probability. There are two simple strategies to reduce the interference – channel allocation and power control.

10.5.2 Channel allocation

An IEEE 802.11 network operating in a 2.5 GHz spectrum (IEEE 802.11b/g) has a total of 11 channels out of which one can have three non-overlapping channels. Figure 10.5 illustrates a channel spectrum for IEEE 802.11b/g. An example of three non-overlapping channels can be 1,6,11 as is also shown in Figure 10.5.

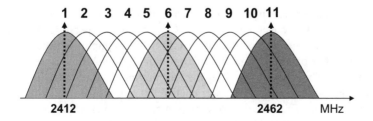

Figure 10.5   Channel spectrum in IEEE 802.11 b/g.

Non-overlapping channels have no co-channel interference. Similarly, an IEEE 802.11a network operating in 5.2 GHz has 12 non-overlapping channels. The objective of channel management is to ensure that there be minimum interference and the overall network capacity be high. For example, if two proximate APs are in the same channel, the channel capacity (11 Mbps for IEEE 802.11b/g) is shared among the APs, thereby lowering the throughput. Also, a transmission on a same channel can result in a hidden terminal effect and in packet collision at the receiver. Packet collisions significantly lower the throughput and also result in unpredictable performance in terms of throughput and delay. Channel management is therefore a necessary and critical component in any IEEE 802.11 wireless network deployment.

Channel management results in an allocation of channels on which each AP should operate. Since the IEEE 802.11 standard does not specify any particular channel allocation algorithm, different vendor-specific algorithms exist. Most of the channel allocation algorithms try to allocate different channels to neighboring APs. For example, if two APs contend with each other or are in the carrier-sense region of each other, the APs are assigned two non-overlapping channels. With the above assignment, each of the two APs can transmit at maximum throughput. Figure 10.6 illustrates a simple example of channel allocation for two specific topologies – *linear* and *triangular*.

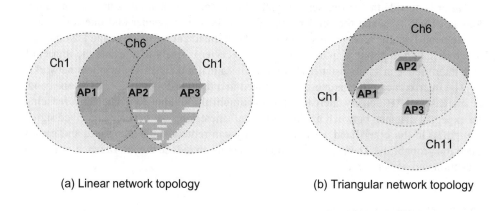

(a) Linear network topology   (b) Triangular network topology

Figure 10.6   Example of channel assignment for two different topologies.

10.5.3 Power control

An AP transmits data at a certain power level. In a typical WLAN deployment, multiple APs operate simultaneously to cover a given space. Clearly, an AP operating at the maximum power level ensures a good coverage with uniform throughput. However, in a dense deployment, a higher operating power results in higher inter-AP interference and lower overall throughput. The process of limiting the transmitting power of an AP is called power control. There are two main goals of power control: (a) to increase spatial reuse and (b) to conserve the battery power of the mobile station. A simple power control approach is to set the maximum transmitting power of an AP to the level that ensures good coverage to the farthest station associated with it. Figure 10.7 illustrates the above case of power allocation.

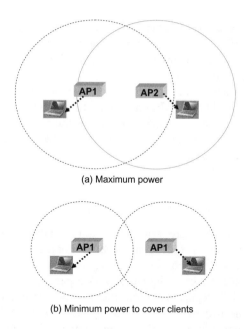

(a) Maximum power

(b) Minimum power to cover clients

Figure 10.7   Example of power allocation for two APs.

In current devices, power can be dynamically adapted based on the position of the stations. In order to conserve battery power at the mobile station, APs provide the functionality of packet buffering for the associated stations. Packet buffering enables the station to enter sleep state when there is no packet transmission intended for the given station. In order to wake up a station when there is data available for that station, the AP sends an advertisement notifying the station for which data is being buffered. The station is required periodically to move into the listen state to receive the advertisement from the AP. Using this mechanism, the station does not waste battery power by remaining in active state all the time.

10.6   IEEE 802.11 STANDARDIZATION OVERVIEW

There are multiple working groups under the IEEE 802.11 charter focusing on defining various functionalities, architectures and extensions in the IEEE 802.11 area. IEEE 802.11e is focusing on the QoS issues that will be discussed in detail in the next chapter. Some of the important working groups relevant to supporting VoIP over WLAN are: IEEE 802.11T, IEEE 802.11e, IEEE 802.11r and IEEE 802.11k.

IEEE 802.11T working group focuses on wireless performance prediction and recommends performance metrics, tests and measurement methods. These recommendations include the QoS requirements for VoIP applications that an IEEE 802.11 product or network should support. IEEE 802.11e is focusing on the provisioning of QoS in IEEE 802.11 networks, which is required for prioritizing VoIP traffic in the presence of data traffic. IEEE 802.11e will be discussed in detail in the next chapter. IEEE 802.11r focuses on providing fast access point hand-offs to ensure seamless mobility. Fast hand-off mechanisms can help in minimizing the data packet loss in a burst during the hand-off – an important requirement in supporting VoIP. IEEE 802.11k focus on radio resource management protocol that can enable the handset to make fast roaming decisions through prediscovering all neighboring APs, their distances and call capacity.

IETF is also standardizing the Control and Provisioning of Wireless Access Points (CAPWAP) protocol to enable thin AP architecture. In such an architecture, all intelligent tasks including client level management is done at a centralized controller, thereby making the AP thin in terms of functionalities. CAPWAP provides a protocol for the AP to communicate with the controller.

10.7 SUMMARY

IEEE 802.11 wireless networks are steadily becoming a popular access technology for carrying wireless data as well as voice for users both at home and in an office environment. Understanding the IEEE 802.11 network technology is thus important for deploying voice, particularly because voice requires a high level of QoS guarantees.

Typically, voice will be deployed in infrastructure configuration with either a wired or multihop mesh backbone. In deployments with a large number of APs for providing wireless coverage to a large region while supporting a large number of clients, both network access and resource management technology is critical. The basic network access technology provides efficient management of association, security and roaming or handoff among APs. Resource management focuses on performance-related issues with the goal of minimizing highly loaded APs, maximizing client throughput and maintaining required QoS.

REFERENCES

1. Gast, M. 802.11 802.11 Wireless Networks: The Definitive Guide, O'Reilly, Sebastopol, CA (2005).

CHAPTER 11

VOICE OVER IEEE 802.11 WIRELESS NETWORKS

IEEE 802.11 wireless networks were originally designed for high-speed data access. While an IEEE 802.11 wireless network can potentially provide high throughput in the absence of interference, the network is not well equipped to meet the QoS requirements of VoIP. Neither is an IEEE 802.11 network designed to differentiate between download traffic and delay sensitive VoIP traffic, nor can it provide the high level of robustness and reliability as required in an enterprise deployment of VoIP over WLAN. In the recent past, a significant amount of development took place in extending the base IEEE 802.11 protocol and proposing new solutions to support VoIP over WLAN. In this chapter, we present the various performance-related challenges and solution approaches to deliver VoIP over WLAN.

11.1 VoIP OVER WLAN PERFORMANCE PROBLEMS

The previous chapter provides a background on deploying and operating an IEEE 802.11-based WLAN system. Once the WLAN is deployed and the wireless connectivity from the handset (Station) to the Internet or Enterprise Network is established, it is easy to deploy a VoIP service. A typical VoIP over WLAN architecture is shown in Figure 11.1. While deployment of VoIP service is simple, meeting the QoS requirements of VoIP is inherently challenging. Following are some of the root causes of performance problems in delivering VoIP over WLAN.

VoIP: Wireless, P2P and New Enterprise Voice over IP   Samrat Ganguly and Sudeept Bhatnagar
© 2008 John Wiley & Sons, Ltd

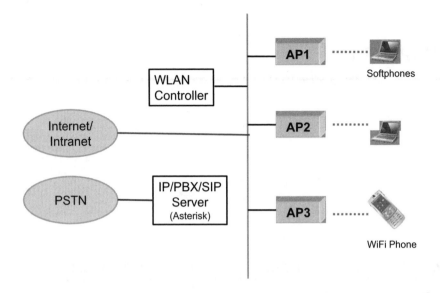

Figure 11.1   VoIP over WLAN architecture.

11.1.1   Channel access delay

An IEEE 802.11-based MAC protocol using Distributed Coordination Function (DCF) was not designed to provide delay guarantees. The characteristic properties of the DCF protocol can result in high variability in the packet transmission delay. The delay is a result of the time taken to access the channel and send the pending VoIP packet to air. The variable delay is caused due to two possible scenarios. First, the MAC waits till the channel becomes idle. This waiting time depends upon the size of the packet currently being transmitted. Second, if there is a collision, the MAC waits for a random amount of time before the VoIP packet is retransmitted. A single VoIP packet can go through multiple such collisions followed by a random back-off and retransmissions. The consequence is that a single VoIP packet can face unpredictable delay in moving from the AP to the client station or vice versa.

11.1.2   Interference from simultaneous transmissions

Ongoing traffic (e.g. HTTP downloads) delivered on the same channel on which the VoIP device is connected to its AP, by neighboring APs or the same AP causes interference. Internal interference from ongoing VoIP calls or data traffic can significantly impact the quality of a given ongoing VoIP call. Active flow of packets from a neighboring AP in same channel can result in packet collisions at the receiver or reduction in the sending rate (as discussed in the previous chapter). Packet collisions at the receiver require the packet to be retransmitted, further adding to the delay. Reduction in the sending rate implies that the packet is queued at the sender and faces additional delay before it can be sent into air. The unpredictable and variable delays from such interference add to significant end-to-end jitter, resulting in performance degradation.

11.1.3   External interference

External interference is caused by the external sources such as microwave, multipath fading effects, etc. External interference reflects in the quality of the wireless medium and is given by signal-to-noise ratio (SINR). While choosing a different modulation technique and reducing the data-rate can increase the SINR, the variability in perceived quality of the medium cannot be avoided. The sudden intermittent periods of interference result in packet retransmissions and associated variation in delays.

11.1.4   Disruption in connectivity

Supporting limited roaming requires handoff across APs. During handoff, the wireless connectivity is broken, resulting in interruption of an ongoing VoIP call. If the handoff duration is short, the packets can be buffered and retransmitted albeit adding to the jitter. However, if the handoff duration is long, a burst of VoIP packets will be lost without being recovered.

11.1.5   Power drain

Typically, a user is expected to use a mobile Wi-Fi battery-operated handset for the VoIP service over WLAN. IEEE 802.11-based wireless access can easily drain the battery due to higher power usage. For example, profile results using Cisco 802.11 cards show that up to 30% of the total power is spent in IEEE 802.11-based wireless communication. Power-drain is a critical factor in the practical deployment of VoIP over WLAN.

11.2   VoIP CAPACITY

Given a deployed IEEE 802.11 wireless network, it is important to understand the VoIP carrying capacity of the network. Accordingly, one can plan how to upgrade the network and deploy the access points. Knowledge of VoIP capacity is required to perform admission control. Typically, the VoIP capacity of a network is defined as the number of VoIP calls that can be supported while meeting the target quality. For example, consider the quality measure as the R-score for the G.729 codec and a target value of 70. Having a VoIP capacity of seven calls would mean that the network can handle seven simultaneous calls where each call's R-score is at least 70.

As we already know, the quality measure for VoIP (R-score or MOS) is a function of the packet loss, delay and jitter. For an ideal network with an error-free channel and no interference, the delay due to queueing makes the maximum impact in degrading the R-score. For example, if the incoming rate of VoIP traffic in packets per second is greater than the network capacity limit, the packets will go through an unbounded queuing delay and the R-score may become as low as 5 or 10. Therefore, the job of network provisioning and call admission control is to ensure that network capacity constraint is met and that there is no queueing delay introduced at any point in the wireless network.

The above discussion amounts to computing the number of calls that a given network can handle without exceeding the capacity. Let us consider a simple scenario to illustrate the approach for computing VoIP capacity. In this scenario, we have VoIP calls using G.729a codec with each packet of length 20 bytes and 20 ms interpacket duration. The 20 ms duration corresponds to 20 ms audio encoded in each packet, which is also the most common usage case. The one-way traffic for this codec amounts to 50 packets/s. Let us consider IEEE 802.11b wireless network operating at 2 Mbps bit-rate consisting of an AP connected to a set of stations.

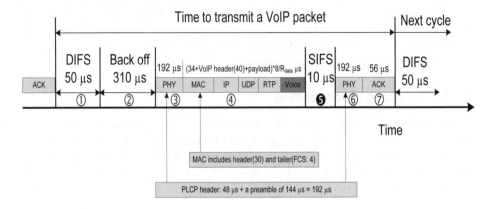

Figure 11.2   MAC overhead in VoIP packet transmission.

The total time the MAC layer takes to send a single voice packet without any retransmission is given in Figure 11.2. The entire process of sending the packet starts with the expiry of DIFS amount of time ($= 50 \ \mu s$) followed by a random back-off time ($\approx 350 \ \mu s$) followed by sending the voice packet into the air. After sending the packet, the MAC layer waits for the acknowledgement (ACK) from the receiver. Once the ACK is received, the process of sending a single voice packet concludes.

One can notice the huge overhead incurred in sending a single voice packet that includes various types of header (including PHY, MAC, IP, UDP, RTP). The total length of the PHY, MAC, IP, UDP and RTP headers is 86 bytes. The total amount of time to send a single voice packet is $1186 \ \mu s$ where the overhead is $946 \ \mu s$ corresponding to around 80% overhead.

The number of calls one can fit in this network is given by the 20 ms/($\times 1186 \ \mu s$), which equals eight VoIP calls. The factor of 2 in the denominator in this computation is due to the fact that the VoIP call is bidirectional where two data streams are established for each call. Interested readers can obtain further details about VoIP capacity over WLAN in [1].

11.2.1   Packet aggregation

From the above discussion on capacity computation, we notice that an IEEE 802.11 wireless network incurs a large overhead in sending a small voice packet. If we extend the above calculations, we observe that the total throughput from eight VoIP calls is 128 Kbps while the total available throughput from the channel is 2 Mbps. The above calculation is for an ideal scenario where we assumed that there are no packet collisions. However, in reality, smaller packets result in a higher probability of collisions and a greater number of retransmissions. Using an actual IEEE 802.11b access point, the number of supported VoIP calls is observed to be six.

A simple yet effective strategy to increase the capacity is to create larger voice packets. This can be done by having the codec digitizing a large amount of audio (50 ms to 100 ms) in each packet. By having higher voice data payload, the overhead from the headers is minimized. For example, a voice packet containing 100 ms of audio is 100 bytes in length. A 100 ms audio packet results in 46% overhead from the headers compared to a 10 ms audio packet incurring 81% overhead from the headers. The resultant effect in terms of capacity is significant. For a IEEE 802.11b network operating at 11 mbps bit-rate, 100 ms audio packets result in 66 supported calls compared to 20 ms audio packets resulting in only 14 supported calls.

In many cases where the audio codec cannot be modified, one can take the approach of performing in-network packet aggregation. In this approach, the multiple voice IP packets are aggregated to create a large IP packet before it is sent to the MAC layer for transmission into air. By having a larger IP packet, the MAC layer overhead from DIFS, SIFS, ACK, retransmissions and back-off is minimized.

The packet aggregation technique is shown in Figure 11.3. The figure illustrates the time that is saved in sending a series of VoIP packets. Implementation of packet aggregation is simple and is done at the IP layer. The technique is implemented by having a packet queue for each destination that receives some of the incoming packets. The packets in a particular destination's queue are aggregated by encapsulating the packet under a new IP header. At destination, the aggregated packets are extracted. It should be noted, the packet aggregation can add delay to each voice packet. For example, a single G.729 voice stream with one packet every 20 ms will add a 100 ms delay if we want to aggregate four packets. For multiple streams, packet aggregation works better as there will exist more packets to aggregate without waiting and adding extra delay. Packet aggregation technique finds its true benefit when applied to VoIP over Mesh Networks [2] (discussed at the end of the chapter).

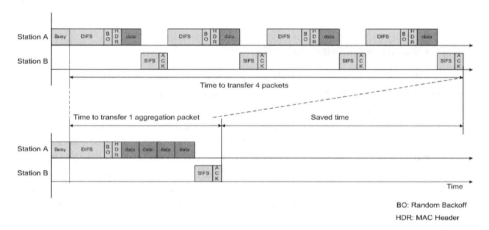

Figure 11.3   Benefit of packet aggregation.

11.2.2   Header compression

Header compression is another technique that can be applied to increase the VoIP capacity and is complementary to the aggregation scheme. The usage for header compression is motivated by two facts: (a) the headers occupy a large portion of the packet (86 bytes for 20 bytes voice packet); and (b) the headers have a significant amount of redundancy.

Packet headers with redundancy may be reduced through compression techniques such as compressed RTP (cRTP) or RObust Header Compression (ROHC) [3]. Typically, ROHC is considered to be the most efficient technique for use over a wireless medium where the bit error rate is high. Using the ROHC technique, the large overhead of 40 bytes per VoIP packet from RTP/UDP/IP can be reduced to 2–3 bytes.

ROHC works by having a compressor and a decompressor at each side of the link. The decompressor stores the connection-ID in terms of IP address and port number in order to refer to the correct VoIP stream. The compression technique acts similarly to video compression where a base packet is sent 'as is' followed by sending only the difference. Therefore, an error in the base packet results in re-synchronizing the states by sending a new packet as the base packet with the original set of headers.

Even with optimizations in ROHC to make resynchronization efficient, high and fluctuating bit error rate can reduce the compression gain. Another complementary approach is to extend the packet aggregation implicitly to perform header compression. It is easy to note that in an aggregate packet, only one component packet is required to carry the entire header. For the rest of the component packets, only the difference can be sent. Such a technique does not require maintaining state synchronization across the link and is not affected by bit-error rate. The reason is that if an aggregated packet is lost, the entire packet is resent automatically.

To provide the reader with a flavor of the capacity improvement with the above techniques, let us consider the G.729a codec with 20 ms audio packet on a 2 Mbps bit-rate link. Without aggregation, the capacity is seven calls. The network capacity increases to 30 calls with packet aggregation and to 71 calls with joint packet aggregation and header compression.

11.2.3 Interference limited capacity

The above discussion considers the VoIP capacity for an isolated access point. However, in a real deployment where there are multiple access points, we need to consider the inter-AP interference in deriving the actual number of VoIP calls that can be supported. As we discussed in the last chapter, if two APs are in the same carrier-sense region of each other, the channel capacity is shared among the two APs. Let us consider these two APs that operate on the same channel. The channel is operated at 2 Mbps bit-rate and can support a maximum of seven VoIP calls. Since these two APs interfere, the total number of supported calls by these two APs combined cannot exceed seven calls at any given time. However, if the two APs do not interfere, the total number of calls that can be supported by these two APs can go up to 14. Therefore, in order to increase the capacity, channel assignment is important to reduce the effect of interference and channel-sharing among access points.

11.2.4 Call admission control

Call admission control should be an important part of deploying VoIP over WLAN to ensure that the network capacity is not exceeded. Based on the codec used and the channel bit-rate, it is easy to figure out the maximum number of calls that can be supported for a given channel. Next, the call admission control algorithm needs to incorporate the inter-AP interference to compute the actual capacity. We next discuss a simple solution for call admission control.

Let us consider a simple scenario where all APs are operating on the same channel that can support a maximum of N calls. The call admission control module resides at the central controller. When a new VoIP call request arrives, the controller checks whether or not the AP that will be used for routing the VoIP call can support it. In order to check, the controller maintains the available capacity for each AP in terms of calls at any given time. When a new call is admitted, the available capacity for a given AP is updated in the following manner. If the new call is routed through the given AP, the available capacity for that AP is decreased by one. If the new call is routed through any AP in the neighborhood that interferes with the given AP, the available capacity is still decreased by one. Note that the above update procedure captures the interference-limited capacity effect. When a call is terminated at the given AP or an interfering AP, the available capacity is increased by one. The controller rejects a call if the available capacity is less than one. Note that the controller may route the call through a different AP where the available capacity is met. The above admission control mechanism is simple to deploy. The limitation of this mechanism is that it does not challenge other effects such as fluctuation in channel condition or hidden terminal problems.

11.3 VoIP PACKET PRIORITIZATION

Existing IEEE 802.11a/b/g based WLAN MAC uses the DCF mechanism for channel access. Please refer to the previous chapter for a detailed discussion on the DCF mechanism. The DCF mechanism achieves fairness among contending packets trying to access the channel and provides a best-effort service to applications. Unfortunately, a best-effort service cannot meet the delay requirement of ongoing VoIP applications. The problem arises in scenarios where an ongoing VoIP traffic is contending with TCP traffic that could be generated from large file download by users. In this scenario, a VoIP packet is queued up at the MAC layer waiting for the time when the channel is free. The waiting time can increase significantly when the network becomes temporarily loaded due to the bursty nature of TCP traffic. Queuing up small VoIP packets behind large TCP packets further adds to the delay. The resulting sudden increase in delay from queuing may cause a series of VoIP packets to miss their deadline and create an intolerable experience. We next discuss some of the practical strategies to prioritize the VoIP flow in a IEEE 802.11 wireless network by studying various cases.

11.3.1 Downlink prioritization

In this case, we consider the most simplistic scenario where the VoIP traffic contends with TCP traffic on the downlink from AP to a set of stations. This simple scenario also assumes that there is only a single AP. The downlink traffic from TCP is typically high since most users download content significantly more than they upload.

Since both the VoIP traffic and TCP traffic enters through the same AP, one can use a dual queue to prioritize the VoIP traffic. The queue can be implemented at the device driver layer without any need to modify the hardware or the firmware and interoperate with any off-the-shelf IEEE 802.11 card. The dual queue strategy maintains one queue for VoIP traffic and another queue for TCP traffic. The incoming packets are classified based on ports or packet length and pushed into the corresponding queue. In order to prioritize the VoIP packets, the packet scheduler operating on the two queues serves the TCP traffic queue only when the VoIP traffic queue is empty.

The above strategy basically overlays an application level queue over the packet level FIFO queue maintained at the device buffer. In order to override the action of the FIFO queue at the device, the size of the device queue is set to one packet. Otherwise, the contention at the device queue will affect the VoIP packet delivery. It is to be noted that to maintain good throughput, typically the device queue length is set to four to five packets instead of to one packet.

The above strategy can implicitly control TCP traffic rate. By setting a limit on the size of the TCP traffic queue, the packets for a TCP session will be dropped when the limit is reached. The drop will make the TCP source back-off and reduce the sending rate. However, it is important to select carefully the maximum size of the queue as otherwise the TCP throughput also can become very low.

11.3.2   Uplink prioritization using IEEE 802.11e

For the uplink, it is difficult to prioritize the VoIP traffic. The potential problem arises where one station is actively participating in a VoIP call while another station engages in uploading a big file using FTP or HTTP. It is difficult to coordinate the actions of the two stations in terms of prioritizing packets of these two applications with the DCF mechanisms of IEEE 802.11a/b/g MAC.

Lack of explicit central coordination was the main factor behind the design of DCF protocol. However, without any central coordination, two packets or two flows cannot be differentiated. IEEE 802.11e has an enhancement to the IEEE 802.11 MAC protocol that is designed to facilitate the differentiation among traffic classes. Recently, IEEE 802.11e is being strongly advocated as the appropriate solution for providing QoS to real-time applications such as VoIP and video streaming. IEEE 802.11e proposes two types of extension: enhanced distributed channel access (an extension to DCF protocol) and hybrid coordination function controlled channel access (an extension to the PCF procotol). We next discuss both the above extensions in IEEE 802.11e protocol and address how the design can help in prioritizing the VoIP traffic at the uplink.

11.3.2.1   Extended distributed channel access (EDCA)   In the original DCF protocol, the station waits for DIFS amount of time to expire before going into the random back-off. In the extended DCF, instead of waiting for the fixed DIFS amount of time, the station waits for Arbitration Inter Frame Space (AIFS) time. The AIFS values are assigned for different priority levels corresponding to application classes such as Voice, Download, Streaming and background. Traffic class with a lower value of AIFS will have to wait for lesser time when the channel is idle compared to the class with a higher value of AIFS. This implies that a VoIP packet will have a lower AIFS and will have more chance to access the medium compared to a TCP packet. Note that the mechanism does not guarantee the exact delay on each VoIP packet, it simply provides a relative prioritization. For example, it is possible that the random back-off timer for the TCP packet expires before that of the VoIP packet giving it a priority.

EDCA also assigns a Transmit Opportunity (TXOP) with each priority level. A TXOP defines a bound on the time interval during which a station can send as many frames as possible that fit the time bound once the station gains access to the channel. The above modification prevents stations operating at low bit-rate to gain access to the channel for an excessive amount of time. Currently, EDCA is becoming the most popular choice supported by the WiFi Alliance for developing the next generation Wi-Fi Multimedia (WMM) standards.

11.3.2.2 *Hybrid coordination function controlled channel access (HCCA)*

HCCA is an extension to the PCF mechanism where there exists a hybrid controller at the AP. The hybrid controller polls stations during a contention-free period. The polling grants a station a specific start time and a maximum transmit duration. Hybrid Coordination Function (HCF) also allows per session control beyond per station control to enable differentiation among traffic classes. The controller can use the feedback on per session level queues to address how the session level queues can be scheduled. Another difference with the PCF is that HCCA assigns TXOP to a station to limit the number of consecutive frames a station can send once it gains access to the channel. HCCA is not currently being considered for deployment.

11.4 HANDOFF PERFORMANCE

In order to maintain the quality of VoIP experience during handoff, the handoff latency needs to be minimized. The entire handoff process can be broken down into two processes: *link-layer handoff*, and *network-layer handoff*. The link-layer handoff consists of the following processes: *probing, authentication* and *association*.

In the link-layer handoff, the probing part is related to figuring out two questions: when to handoff to a new AP, and to which AP to handoff. Once the new AP is selected and the client has decided to handoff to the new AP, the client goes through authentication and then completes the association process with the new AP. The details of the authentication and association processes are discussed in the previous chapter.

The task of the Network-layer handoff is to ensure that frames are correctly delivered to the client when its point of attachment to the AP has undergone a change. The network-layer handoff manages the routing table updates and can also handle interoperability among different APs through the Inter-Access Point Protocol or IAPP (IEEE 802.11f). The network layer handoff can also aid in caching packets at old APs and transferring them to the new AP after the client is associated to the new AP.

We observe that the entire hand-off process goes through multiple operations. Each of these steps adds to the latency during which a client experiences a complete gap in connectivity. In minimizing the handoff latency, it is now important to understand the relative delay added by each of these steps towards the overall latency. Table 11.1 below shows a typical delay estimate for each step based on empirical results from experiments. Note that there exists a large difference in the actual observed latency depending upon the hardware or the card manufacturer used.

Note that the delay in probing is the dominant factor in increasing the handoff latency. Probing delay contributes approximately 70–80% to the total handoff latency. Given this significant delay in probing, we focus our discussion on the probing process and different strategies for minimizing the probing delay.

11.4.1 Probing process

When probing, the client goes through explicit scanning of each channel to find available APs on each channel. There are two scanning approaches: *passive* and *active*. In passive

Table 11.1   Latency of different steps in the handoff process.

Steps	Latency (ms)
Probing	200–400
Authentication	≤ 10
Deauthentication	10–20
Association	≤ 10
Network-layer handoff	13–25

scanning, each AP advertises itself through periodic beacon messages. The client has to wait until it receives a beacon message to discover the existence of the corresponding AP. In active scanning, the client issues an active probe and waits for the response from the AP to discover its existence. In passive scanning, the total worst case delay is of the order of $C \times T$ where C is the total number of channels scanned and T is the duration between the beacon messages. Typically, T is around 100 ms. In the case of active scanning, the time spent on each channel is based on the amount of waiting time to collect a response from all APs. This time depends on the number of APs available for a given channel and can be around 50–60 ms.

Clearly, scanning 11 channels for the IEEE 802.11b network can result in a worst case delay of more than 1 s. More than 200 ms delay in VoIP results in packet loss and significant degradation of voice quality.

Another important factor affecting the performance is the handoff triggering decision. The handoff process is triggered when the client observes packet losses or a significant decrease in the SNR so that it falls below a given threshold. If the triggering decision is too optimistic, there will be too many handoffs. A pessimistic decision will result in packet losses and degradation in VoIP quality. Therefore, an efficient triggering decision is also a critical part of the probing process.

11.4.2   Scanning using neighbor graph

One approach to reducing the cost of probing is to limit its scope. The idea is to not have the device undergo a brute force scan of all possible channels and all possible APs and to limit the scan to a candidate set of a few channels and APs. Minimizing the scanning time by limiting the candidate set is done by maintaining a neighbor graph – an approach proposed in [4]. In this approach, for each AP, a list of neighboring APs for different channels is maintained. The neighbor graph is a static graph that can be created by analyzing the AP location and coverage or can be generated by analyzing the mobility pattern of the users.

For example, consider collecting a mobility history for all clients for a given AP. For each client, the history lists the neighboring APs and the corresponding channel on which it connected after hand-off. This type of history is collected for all clients of all APs. From this history, a simple neighbor graph can be constructed for each AP, listing its neighboring APs and their corresponding channels. Using this graph representation, the clients have to send probe messages only to the neighboring APs. The approach is illustrated in Figure 11.4(a). Note that, the number of channels to scan through is also minimized. The probing latency can be reduced to less than 50 ms by using the neighbor graph information.

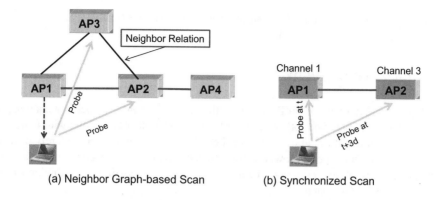

(a) Neighbor Graph-based Scan   (b) Synchronized Scan

Figure 11.4   Probing strategies in handoff.

11.4.3   Synchronized scanning

Another approach proposed in [5] to minimize the delay is by synchronizing the APs in sending their beacon messages. In this approach, an AP operating in channel i sends a beacon message at time $t + id$ where t is the start of the period and d is a constant duration between beacons sent on successive channels. The approach is illustrated in Figure 11.4(b). If a client knows this schedule, it can shift to the corresponding channel and obtain the required information from the APs operating on that channel. The above synchronous approach can be done in a proactive way, where the client continuously shifts through channels and discovers the APs and their signal strengths. Based on the information, the client can make a decision at any point to time to undergo handoff. At handoff, the client does not have to initiate another scanning and instead can skip to the authentication and association step. The continuous probing can be done every 500 ms without adding too much overhead from the proactive probing. The above approach can work complementarily to the neighbor graph approach.

11.4.4   Multiscanning using dual radio

By leveraging two radio interfaces at the client station, the handoff latency can be reduced to almost zero. The availability of multiple radios at the client is uncommon but not improbable. Many manufacturers are providing multiple radios to boost throughput (e.g. www.engim.com).

In the multi-scanning approach, a client station has a primary interface and a secondary interface. While the primary interface is busy communicating with AP, the secondary interface can be used for scanning. Once the scanning result is obtained, the primary interface can hand off to the new AP bypassing the scanning process altogether and reducing the latency to less than 40 ms. Another possibility is to have the secondary interface go ahead with the entire handoff process, completing association with the new AP. In this case, the latency is almost zero. Once the secondary interface is switched to the new AP, the client station makes the secondary interface primary, and vice versa. One of the issues in using multiple interfaces is excessive use of battery power. Therefore, the secondary interface should be used in an efficient manner for the scanning purpose. Further details can be found in [6].

11.5 RELIABLE DELIVERY

Assuring reliable delivery of VoIP packets in the presence of interference is a challenging task. Interference either from other traffic or from an external source along with multipath fading effect can result in bursty loss of packets. Bursty packet loss can severely degrade the quality of the user experience where, for instance, a part of the speech may be lost. Such bursty packet loss is quite common in current IEEE 802.11 networks operating on shared channels in the 2.4 Ghz band.

A simple yet effective approach to relieve bursty packet loss is proposed in the Divert solution [7]. In this solution, there exists a Divert controller through which the packets are routed to the APs and subsequently to the client. Each AP monitors the link quality and sends it as feedback information to the Divert controller. If the Divert controller finds that the link is experiencing interference as a result of a low delivery ratio, the controller diverts the traffic through another AP. The approach is illustrated in Figure 11.5(a). The switching is done at real-time and can have a fast reaction to the interference effects. Experimental results demonstrate that operating under normal condition, Divert can reduce the frame error rate by 26%.

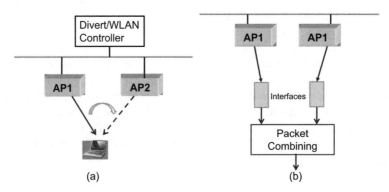

(a)   (b)

Figure 11.5   Example of channel assignment for two different topologies.

Just as we discussed in the hand-off solution, one can leverage multiple radios at a client station to provide reliable delivery of VoIP packets. In this solution proposed in [8], the same packet is sent from more than one AP to the client station and vice versa by using each radio to connect to a unique AP as illustrated in Figure 11.5(b). In the downlink path, the station receives a copy of the packet from each interface, and selects the packet that is error-free. A packet is considered error-free if it passes through the cyclic redundancy checksum (CRC) check. If neither interface provides an error-free packet, a block-based combining algorithm is employed to construct an error-free packet from the set of corrupted packets. In order to implement the solution, the client station is required to have a packet buffer that ensures in-order delivery of packets and no duplicate packet forwarding. The solution can be extended to the uplink side by performing the same operation at the controller through which all the packets are routed to the outside network. The total delay incurred in the packet combining process is less than 35 ms for which a significant reliability improvement is observed in the experimental study.

11.6 CLIENT POWER MANAGEMENT

Current IEEE 802.11 cards provide a power save mode of operation significantly to reduce the power consumption. If this mode is used, the client station notifies the AP that it is going to the *sleep* state. The AP starts buffering the packets intended for the client. The client periodically wakes up to listen to the beacon from the AP. Using the beacon, the AP notifies the given client whether any packets are buffered for the given client. Based on this notification, the client wakes up and starts receiving packet transmission from the AP. In an alternative option, the client station wakes up periodically and sends a PS-Polls request message to the AP. Receiving this message, the AP sends out any buffered data to the client station. By completely turning off the entire RX/TX circuitry, the client can save considerable power using the power save mode.

Unfortunately, the power save mode is not appropriate for VoIP traffic since buffering packets can result in significant latency. Furthermore, the data exchanged in terms of notifications can add a significant overhead for VoIP traffic with 20 bytes packet length. In order to alleviate the problem in supporting VoIP, the IEEE 802.11e standard includes an optional extension of the standard power save mode defined as Automatic Power Save Delivery (APSD). The main difference in the APSD mode is that now the station is awake during the service period as opposed to being awakened as triggered by the beacon or listen intervals expiry.

There are two types of Service periods under the APSD mode: *Unscheduled* and *Scheduled*. In Unscheduled APSD or U-APSD [9], the uplink data from the station is used as an indication that the station is in active state. This indication triggers the AP to send any buffered frame back to the station. The service period in U-APSD thus begins with the trigger from the uplink data and ends when the AP sends a notification for End-of-Service-Period referring to 'no more data to be sent'. The U-APSD is particularly suitable for bidirectional VoIP traffic, where the data rate is the same in both directions. Using U-APSD, the station enters into a doze state until the next voice data has to be sent to the AP.

In scheduled APSD, a pre-arranged wake time for each of the stations is notified by the AP. This allows different stations to wake up at different times. Schedule APSD is therefore efficient for supporting multiple VoIP calls and user density. Scheduled APSD is also more appropriate for VoIP since the voice packets are sent and received after fixed intervals. However, unlike Scheduled APSD that is meant only for streaming data, U-APSD can support both voice and data.

11.7 ISSUES IN MESH NETWORKS

IEEE 802.11-based wireless mesh networks is considered as an inexpensive solution for building and deploying the wireless backhaul network that serves a set of geographically distributed APs. With a wireless mesh network, the VoIP traffic has to follow a multihop path to the client station. The multihop path adds to several performance related problems in supporting VoIP. In this section, we limit our discussion to a few sets of problems related to capacity and reliable delivery of VoIP over Mesh Network. For further reading, see [2].

11.7.1 Capacity in mesh networks

In order to understand the fundamental issues arising in the mesh networks, it is important to understand the effect of the number of wireless hops on the number of supported VoIP calls. With the increase in the number of hops, the capacity decreases. Figure 11.6 (without aggregation) shows the effect for a linear topology for G.729A codec with each link operating at 2 Mbps bit-rate. The decrease in capacity on multiple hops is due to the self-interference effects. For example, with two hops, the capacity is halved since the node in the middle cannot receive and send at the same time. Therefore, the forwarding of data by the middle node happens at half speed. Also multiple nodes on a given path may be interfering. This implies that at given time only one of these sets of nodes can transmit. The above problems limit the diameter in terms of number of hops for a mesh network.

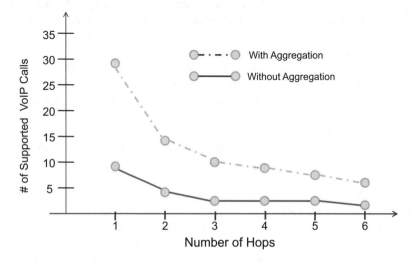

Figure 11.6   VoIP capacity in mesh network with and without aggregation.

Given this multihop effect on capacity, it makes more sense in a mesh network to apply solutions for increasing capacity. As we discussed earlier, packet aggregation can help significantly in terms of improving capacity. Figure 11.6 shows the increase in capacity with packet aggregation. Since a mesh network acts as a wireless backhaul network serving multiple APs, multiple calls routed over a path to different APs create better opportunities for packet aggregation.

11.7.2 VoIP call routing

Routing VoIP call relates to discovering the end-to-end multihop path from the gateway to the AP. Typically, the best way is to have some proactive approach for routing the VoIP calls. A proactive approach by having routes prediscovered avoids the delay in routing calls when a new call arrives. Pre-discovered routes can be pinned down using label-based forwarding techniques, allowing an even faster approach to routing new calls.

The exact path to route the calls depends on various factors. One approach is to route the calls through links with high delivery ratios (links where packet loss is minimum). However, route fluctuation due to time-varying changes in link metrics may negatively impact the experienced VoIP quality. During route changes, packets may become lost, so any routing protocol using link metrics needs a high level of hystersis to avoid route fluctuation.

In many cases, the link metrics are stable and good. In such cases, the routing decision can be based on the end-to-end capacity of the path. For example, if a particular link is loaded, the route of the new call should avoid the link.

11.8 SUMMARY

With the proliferation of IEEE 802.11 devices, VoIP over IEEE 802.11 is becoming a natural choice for many enterprise and home users. However, the adoption of the VoIP service will see a widespread adoption only when the user experience is satisfactory. To that end the WMM and IEEE 802.11e standards body are working to solve the QoS issues related to supporting VoIP service. In going forward towards deploying VoIP over IEEE 802.11, the main issues to address are:

- increasing capacity on limited and shared bandwidth IEEE 802.11 network;

- reliability in Voice packet delivery;

- power drain factor.

At the same time, VoIP service provider or operator cannot ignore the new mesh network technologies for extending VoIP service beyond just hotspots. To deploying VoIP over Mesh Networks, a new set of performance problems that originate due to the multihop packet routing feature need to be addressed.

REFERENCES

1. Hole, D. and Tobagi, F. Capacity of an IEEE 802.11b Wireless LAN supporting VoIP, *IEEE Conference on Communications* (2004).

2. Ganguly, S., Navda, V., Kim, K., Kashyap, A., Niculescu, D., Izmailov, R., Hong, S. and Das, S. Performance optimizations for deploying VoIP services in mesh networks, *IEEE Journal on Selected Areas in Communications (JSAC) Multi-hop Wireless Mesh Networks* 4th Quarter (2006).

3. Borman, C. *et al.* RObust Header Compression (ROHC): Framework and four profiles: RTP, UDP, ESP, and uncompressed, RFC3095 (2001).

4. Shin, M., Mishra, A. and Arbaugh, W. Improving the latency of 802.11 hand-offs using neighbor graphs, *ACM/USENIX Mobisys* (2004).

5. Ramani, I. and Savage, S. SyncScan: practical fast handoff for 802.11 infrastructure networks, *IEEE Infocom* (2005).

6. Brik, V., Mishra, A. and Banerjee, S. Eliminating handoff latencies in 802.11 WLANs using multiple radios: Applications, experience, and evaluation, *ACM/USENIC Internet Measurement Conference* (2005).

7. Miu, A., Tan, G., Balakrishnan, H. and Apostopoulos, J. Divert: Fine-grained path selection for Wireless LANs, *ACM Mobisys* (2004).

8. Miu, A., Balakrishnan,H. and Koksal, C. Improving loss resilience with multi-radio diversity in wireless networks, *ACM Mobicom* (2005).

9. Chen, Y., Smavatkul, N. and Emeott, S. Power management for VoIP over IEEE 802.11 WLAN, *IEEE Wireless Communication and Networking Conference* (2004).

CHAPTER 12

IEEE 802.16 WiMAX

IEEE 802.16 WiMAX is gaining significant acceptance as a future wireless technology driving wide-area *Broadband Wireless Access (BWA)* to users. By being inherently all-IP and packet switched technology, unlike existing cellular technologies, WiMAX can directly support VoIP. WiMAX also brings various advantages over competing third-generation cellular technologies. These includes higher spectrum efficiency, higher rate, better QoS management, ease of deployment and interoperability with existing IP networks. Existing deployments of WiMAX in several countries to provide broadband access already demonstrate the successful adoption of WiMAX. This chapter focuses on providing a preliminary background on the underlying IEEE 802.16-based WiMAX technology to understand the full potential of WiMAX as the future wireless broadband technology to support VoIP.

12.1 WiMAX OVERVIEW

WiMAX stands for Worldwide Interoperability for Microwave Access. WiMAX is based on IEEE 802.16 wireless MAN standard. Typically, WiMAX equipments are designed to operate in the 2.5 GHz, 3.5 GHz and 5.7 GHz frequency bands. The underlying IEEE 802.16 standard uses OFDM physical layer technology. WiMAX is supposed to provide a potential data rate of up to 70 Mbps. However, at present most of the reports indicate that the actual measured data rate is of the order of 10 Mbps. WiMAX is supposed to provide a

VoIP: Wireless, P2P and New Enterprise Voice over IP   Samrat Ganguly and Sudeept Bhatnagar
© 2008 John Wiley & Sons, Ltd

larger coverage range of up to 20 miles. This has enabled it to be positioned as a provider of last mile broadband access as an alternative to cable and DSL technologies.

Furthermore, ongoing standardization of WiMAX under IEEE 802.16e (referred to as Mobile WiMAX) is enhancing WiMAX technology to support mobility as well. Mobile WiMAX is based on using *Orthogonal Frequency Division Multiple Access* (OFDMA) as the physical layer technology to support multiple users in a scalable way. Mobile WiMAX-based networks are considered as a potential alternative to the traditional cellular networks in providing connectivity to mobile users by allowing handoff among base stations.

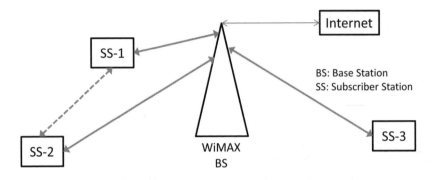

Figure 12.1   WiMAX network topology.

A simple form of WiMAX-based network architecture is shown in Figure 12.1. An important point to note from the figure is that WiMAX supports two configurations: PMP and Mesh. PMP stands for *Point to Multipoint* where a single base station (BS) serves multiple subscriber stations (SS). Mesh configuration allows two SSs to communicate with each other.

We next describe the core ingredients of a WiMAX technology. These include the MAC protocol and the underlying OFDM-based physical layer. The focus is limited to technologies that differentiate WiMAX from competing technologies, particularly in terms of supporting high data rate and QoS-constrained applications such as VoIP.

12.2 IEEE 802.11 MAC PROTOCOL ARCHITECTURE

The MAC layer of IEEE 802.16 is comprised of three sublayers which interact with each other through the *Service Access Points* (SAPs), as shown in Figure 12.2. The service-specific convergence sublayer provides the transformation or mapping of external network data with the help of the SAP. The MAC common part sublayer receives this information in the form of *MAC Service Data Units* (MSDUs), which are packed into the payload fields to form *MAC Protocol Data Units* (MPDUs). The privacy sublayer provides authentication, secure key exchange, and encryption on the MPDUs and passes them over to the physical layer. Of the three sublayers, the common part sublayer is the core functional layer which performs bandwidth allocation and establishes and maintains connection. Moreover, as the IEEE 802.16 MAC provides a connection-oriented service to the subscriber stations, the common part sublayer also provides a *Connection Identifier* (CID) to identify which connection the MPDU is servicing.

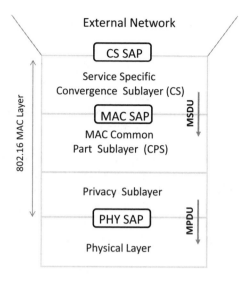

Figure 12.2   WiMAX MAC layer interaction. Reproduced by permission of © 2004 IEEE.

The connection-oriented service is created by defining two concepts: *Connections* and *Service Flow*. Connections refers to a unidirectional MAC layer connection between Base Station and Subscriber Station. Connections are identified by CID. A service flow refers to the MAC layer transport service that maps packets to a CID. A service flow is characterized by a set of QoS parameters. A service flow is identified by a *Service Flow Identifier* (SFID).

12.2.1   QoS management

The above definition of connections and service flows makes QoS management and provisions in IEEE 802.16 easy and flexible. Service flows can be considered as identifiers of application type. For example, all VoIP traffic will belong to the same service flow. Therefore, a service flow corresponds to a set of QoS parameters applicable for a given application. IEEE 802.16 defines an extensive set of QoS parameters. Some of these parameters are defined below.

- *Scheduling Service Type:* This defines the packet scheduling policies such as best effort, real-time polling service, non real-time packet polling service, etc.

- *Traffic Priority:* This defines the priority level for the traffic. A higher priority will result in lower delay.

- *Maximum Traffic Rate:* This defines the peak traffic rate that is allowed.

- *Maximum Traffic Burst:* This defines the maximum burst size that is allowed with reference to physical layer resource allocation.

- *Minimum Reserved Rate:* This defines the minimum rate reserved for this service flow.

- *Tolerate Jitter:* This defines the maximum allowable jitter for this service flow.

- *Maximum Latency:* This defines the maximum allowable latency between BS and SS.

A service flow can be of three types: *provisioned, admitted* and *active*. A provisioned service flow does not have any resources booked for it. An admitted service flow implies that it is feasible to admit the flow but actual resources are not committed to it. An active service flow refers to a state where the resources corresponding to the QoS parameters for the flow are committed by the BS. A connection is always associated with an admitted or active service flow.

In order to provide application level QoS, the incoming packets are first classified using a packet classifier and mapped to a given CID. Each CID is also associated with a SFID. For the given CID, the SFID will provide the set of service-related QoS parameters. Therefore, it is easy to identify each application and apply corresponding QoS constraints to the connection specific transport of the application's packets. WiMAX can also handle QoS-based differentiation between, say, VoIP and data traffic (e-mail, web browsing, etc).

		Bandwidth		Delay		Jitter	
Class	Application	Guideline		Guideline		Guideline	
1	Interactive Gaming	Low Bandwidth	50 Kb/s	Low Latency	80 ms		
2	VoIP	Low Bandwidth	32–64 Kb/s	Low Latency	160 ms	Low Jittering	<50 ms
3	Streaming Media	Moderate to High Bandwidth	<2 Mb/s			Low Jittering	<100 ms
4	Instant messaging/ Web Browsing	Moderate Bandwidth	2 Mb/s				
5	Media Content Download	High Bandwidth	10 Mb/s				

Figure 12.3 QoS guidelines for five application classes.

Based on the above functionalities and provisions, the WiMAX standard defines five classes of application and a set of guidelines in terms of the QoS requirements. Figure 12.3 lists the set of guidelines for these application classes.

12.3 MAC LAYER FRAMING

In the IEEE 802.16 standard, a MAC frame is referred as MAC PDU. The general structure of a MAC PDU is shown in Figure 12.4. A MAC PDU consists of three parts:

- MAC Header (6 bytes) consisting of a generic MAC frame header (GMH) and optional subheaders;

- Payload;

- CRC (4 bytes).

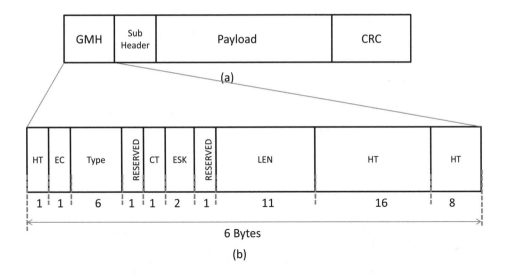

Figure 12.4   (a) MAC PDU; (b) general MAC header.

When the HT bit at the beginning is set to 0, it indicates that the header is a GMH. The EC bit indicates whether or not the payload is encrypted and, if so, the encryption key sequence (EKS) bits indicate which key was used to encrypt the frame payload. The 'Type' field reflects the content of payload in terms of whether aggregation, fragmentation, Automatic Repeat Request (ARQ), or mesh feature of the MAC is used. CI bit, when set, reveals the presence of error-correction codes at the end. The 'LEN' field indicates the number of bytes in the MPDU including the header and the CRC. The CID defines the connection that the packet is servicing. A Header Check Sequence (HCS) is appended at the end of GMH, which works as the cyclic redundancy code for the GMH. The optional subheaders are used to define the bits necessary for aggregation, fragmentation, ARQ and mesh features of the MAC.

The MAC PDU payload consists of zero or more complete MAC SDU or fragments of MAC SDU. The MAC common part sublayer receives information from the upper layer as MAC SDU, which is then packed into forming MAC PDU before being transmitted on a connection. The MAC common part sublayer provides three sets of functionalities in forming MAC PDUs from a set of MAC SDUs as described next.

12.3.1 Aggregation

The common part sublayer is capable of packing more than one complete or partial MSDUs into one MPDU. In Figure 12.5(a), we show how the payload of the MPDU can accommodate more than two complete MSDUs, but not three. Therefore, a part of the third MSDU is packed with the previous two MSDUs to fill up the remaining payload field, preventing wastage of resources. The payload size is determined by on-air timing slots and feedback received from the subscriber station. The decision to aggregate frames is made by the transmitting station. Aggregation is particularly useful when the application layer packets are small such as in VoIP (as was discussed in the last chapter). However, with aggregation, a loss of a single packet results in the loss of multiple MAC SDUs.

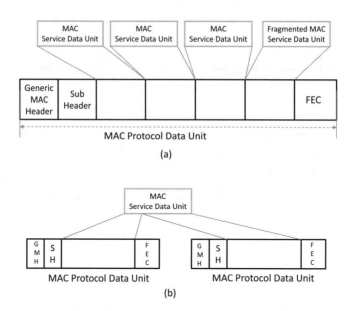

Figure 12.5   (a) Aggregation; (b) fragmentation.

12.3.2   Fragmentation

The common part sublayer can also fragment an MSDU into multiple MPDUs. In Figure 12.5(b), we show how a portion of a single MSDU occupies the entire payload field of an MPDU. Here, the payload field of the MAC packet data unit to be transmitted is too small to accommodate a complete MSDU. In that case, we fragment a single MSDU and pack the fragmented part into the payload field of the MPDU. Fragmentation is useful when the channel is bad and loss of an MPDU does not result in a complete loss of the MSDU.

12.3.3   Concatenation

Through this functionality, multiple MAC PDUs are concatenated into a single transmission. In other words, multiple MPDUs of different CIDs are all grouped into the same physical layer burst.

12.4   PHYSICAL LAYER

The distinctive and most critical requirement of the IEEE 802.16 PHY is that it has to provide high performance while keeping the complexity low. Applications such as VoIP require flexibility for downstream transmission with support for a number of users with possibly variable throughput requirements. IEEE 802.16 also supports multiple access on the upstream transmission. Multiple carrier modulation is beneficial in this regard as it enables the signals in both frequency and time domains to be controlled. Thus, the IEEE 802.16 standard is based on OFDM which was selected in preference to the competing techniques such as single-carrier (SC) and CDMA owing to its superior non line of sight

(NLOS) performance. This permits significant equalizer design simplification to support an operation in multipath propagation environments.

The IEEE 802.16 PHY also provides greater flexibility in terms of modulation and coding as SSs may be located at various distances from the BS and hence may experience different SNR ratios. The BS dynamically adjusts the bandwidth, modulation and coding schemes to overcome the varying SNRs and provides improved system performance. OFDM coupled with forward error correction (FEC) techniques, such as ReedSolomon and convolutional coding, is used when implementing the OFDM PHY. We discuss the underlying OFDM technology in more detail next.

12.4.1 OFDM

In a simple frequency division multiplexing setting, a single communication refers to a single carrier over which data is modulated and transmitted from sender to receiver. Instead of using the entire channel spectrum through a single carrier, OFDM employs the use of multiple subcarriers. Subcarriers are chosen in such a way that even though the corresponding subchannels overlap, but they do not interfere with each other. In other words, the peak of one subchannel always falls on the null point of the overlapping subchannel. A simple way to understand OFDM is to take it as breaking a given channel into many small (narrowband) subchannels. Input symbols from the data are transmitted in parallel using multiple subchannels. As with typical FDM, in OFDM, each subcarrier is modulated with a conventional modulation scheme such as BPSK, QPSK, QAM. However, due to using narrow band subchannels, the modulation is done at a lower symbol rate.

OFDM technology has multiple advantages that cumulatively result in higher spectrum efficiency. OFDM provides better adaptation to channel condition and robustness against intersymbol interference and fading caused by multipath propagation. Lower intersymbol interference is a result of making use of a guard interval between symbols. By having many subchannels, it is easy to introduce guard intervals.

The primary advantage of OFDM over single-carrier schemes is its ability to cope with severe channel conditions – for example, attenuation of high frequencies in a long copper wire, narrowband interference and frequency-selective fading due to multipath – without complex equalization filters. Channel equalization is simplified because OFDM may be viewed as using many slowly modulated narrowband signals rather than one rapidly modulated wideband signal. Low symbol rate makes the use of a guard interval between symbols affordable, making it possible to handle time-spreading and eliminate intersymbol interference.

12.4.2 OFDMA

OFDMA extends the concept of subchannels using multiple subcarriers to support the different rate requirements of multiple users – a typical scenario in point-to-multipoint WiMAX deployment. A subchannel in OFDMA is a logical channel defined as a subset of subcarriers. Typically, one or more subchannels are used to serve a given user. Subchannels can be formed by using adjacent or non-adjacent subcarriers as shown in Figure 12.6. The formation of subchannels allows simultaneous data transmission to/from several users.

OFDMA also allows the number and choice of subcarriers for each subchannel to be changed. This flexibility results in multiple advantages in terms of dynamic resource allocation and higher spectral efficiency. We discuss these features in detail next.

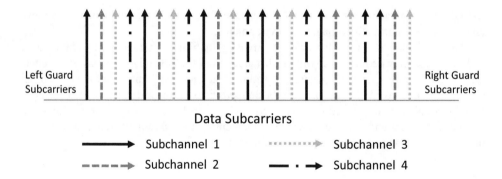

Data Subcarriers

→ Subchannel 1   ▷ Subchannel 3

---→ Subchannel 2   —·→ Subchannel 4

Figure 12.6   Subchannel formation from subcarriers.

12.4.3 Slotted allocation

For dynamic resource allocation, a slot or the minimum data unit needs to be defined. A slot in OFDMA is defined in two dimensions: time and subchannel. For example, as shown in Figure 12.7(a), a slot is defined as three subchannels and two symbols. In a simple sense, each symbol refers to unit time in the time axis. Using the concept of slots, a data burst is a two-dimensional allocation of a group of slots. Different users may be allocated different bursts as shown in Figure 12.7(b).

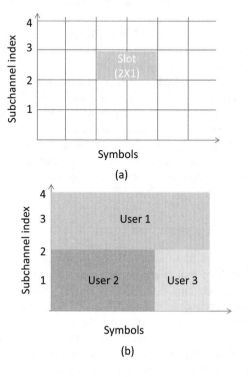

Figure 12.7   (a) Slot formation; (b) user allocation of data bursts.

The above allocation principle suggests two important properties: the minimum allocation of resource is in terms of single subchannel and not subcarriers; and there is no particular restriction on how subcarriers are mapped to subchannels. The second property is an important one and provides two different choices depending on the mode of operation. This aspect is discussed next.

12.4.4 Subcarrier mapping

There exist two ways of mapping subcarriers to subchannels: distributed and adjacent. In the distributed mapping, the subcarriers are mapped pseudo-randomly to the subchannels. This mapping rule has the following advantages: better exploitation of frequency level diversity; and lower probability of intercell interference. In particular, a random distribution results in lower probability that that same subcarrier is assigned to the adjacent cell. In adjacent mapping, adjacent subcarriers are mapped to a given subchannel. Since, the mapping is more restrictive and fixed, it is easy to estimate the subchannel condition and perform adaptive modulation and coding. The adjacent mapping rule is particularly suitable for fixed connections.

The above distributed mapping rule is used under two different modes: Full Usage of Sub-Channels (FUSC) and Partial Usage of Sub-Channels (PUSC). The FUSC mode utilizes or maps the entire subcarriers set to subchannels. The FUSC mode is used only for downlinks. The PUSC mode utilizes a partial set of subcarriers to form subchannels and is used for both uplinks and downlinks. Following are some of the typical structures of each mode.

- *PUSC (DL):* In downlink PUSC, a total of 30 subchannels are available for each symbol. Each subchannel consist of 24 data subcarriers. Each slot is defined as one subchannel and two symbols, which results in a total of 48 subcarriers used per slot. A data region is specified by using the following parameters: subchannel offset, OFDMA symbol offset, number of OFDMA symbols and number of subchannels.

- *PUSC (UL):* PUSC uplink is the same as PUSC downlink, except that each slot is made of one subchannel and three symbols.

- *FUSC:* In FUSC mode, each symbol can use 16 subchannels and each subchannel consists of 48 subcarriers. In this mode, the data subcarriers are first divided into groups of adjacent subcarriers. Then, each subchannel is constructed using subcarriers from each group.

12.4.5 OFDMA frame structure

A typical OFDMA frame structure for downlink and uplink is shown in Figure 12.8. The Downlink Frame contains UL-MAP and DL-MAP. These maps contain information about each data burst in terms of subchannel offset, OFDMA symbol offset, number of OFDMA symbols and number of subchannels. The ranging part contains uplink scheduling information.

12.4.6 OFDMA MIMO

The MIMO system uses multiple antennas both at the sender and at the transmitter to create multiple paths over which information can be sent simultaneously. Multiple uncorrelated

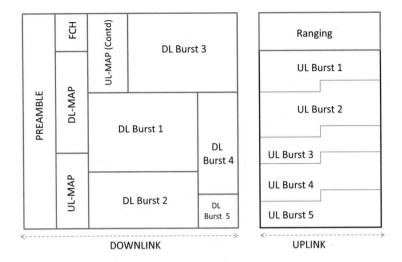

Figure 12.8   OFDMA frame structure.

paths or spatial diversity can be exploited either to increase the reliability of a channel or to increase the throughput. The MIMO technique is used with OFDM to provide a higher rate with better spectral efficiency. With MIMO technology, the multiple simultaneous symbols generated by OFDMA can be sent over spatially diverse paths resulting in a substantial increase in the achieved throughput.

12.5   RADIO RESOURCE MANAGEMENT

In the previous section, the basics of the underlying physical layer technology were described. We next discuss how the radio resources are allocated and managed among different connections. First, it is important to understand how the radio resources are shared and allocated among the uplink and downlink connections.

12.5.1   Duplex modes

There are two types of duplex mode for multiplexing uplink and downlink connections onto the physical layer link: Frequency Division Duplexing (FDD) and Time Division Duplexing (TDD). In the FDD mode, uplink and downlink connections are on separate frequencies. In the case of TDD, both uplink and downlink connections share the same frequency spectrum but are separated in time. The main advantage of TDD is that asymmetric connections can be easily supported.

12.5.2   Uplink bandwidth allocation

The downlink bandwidth allocation is fairly straightforward as it is locally managed by the BS. For uplink bandwidth allocation, there are four different methods:

1. *Unsolicited Bandwidth Grants:* In this type of bandwidth allocation, slots are reserved for the corresponding SS. This method is useful for applications requiring a fixed rate such as VoIP.

2. *Piggyback Bandwidth Requests:* In this method, the request for bandwidth allocation is piggy-backed over an existing MAC PDU without creating a complete MAC PDU only for the request.

3. *Unicast Polling:* In unicast polling, each SS is individually polled by the BS. The polling process is implicit, where uplinks request slots are reserved. Using these slots, each SS sends their bandwidth requests. Unicast polling is more appropriate than unsolicited grants in cases where the bandwidth demand is not known.

4. *Multicast Polling:* When there are many users, unicast polling may result in a high overheard from bandwidth requests. In such a scenario, there is a provision for group-based polling that is also referred to as broadcast or multicast polling. In this case, the uplink bandwidth request slot is not defined for each individual SS but for a multicast or a broadcast group. The SS belonging to the group uses this slot to send the bandwidth request. However, multiple SS sending bandwidth requests may result in collision. In order to resolve collision, the standard specifies a collision resolution rule using random back-off. Using a random backoff timer, an SS decides which slots to use for sending its bandwidth request.

These methods essentially address the problem of uplink bandwidth allocation in a large variety of scenarios.

12.6   COMPETING TECHNOLOGIES

Currently there are three deployment-ready wireless technologies for broadband wide-area data access. They are IEEE 802.11-based WLAN, 3G Cellular data access and IEEE 802.16-based WMAN/broadband access. At the same time, there are other deployment (not ready yet) technologies in the horizon as part of the 3GPP and 3GPP2 cellular technology evolution. They are Long-Term Evolution (LTE) from 3GPP and Ultra Mobile Broadband (UMB) from 3GPP2.

12.6.1   Comparison with IEEE 802.11 WLAN

Both are inherently all-IP based and support high bandwidth. The main difference is in the QoS management area. The base IEEE 802.11 WLAN only supports best effort services while IEEE 802.16 supports multiple QoS classes. Lack of QoS in WLAN is due to the CSMA/CA-based multiple access technology. In contrast, IEEE 802.16 allows TDMA multiple access that allows for exact bandwidth reservation. However, enhancements of IEEE 802.11 as proposed in IEEE 802.11e allow for some level of data prioritization albeit lacking strict QoS guarantees. The second difference between the two technologies is that an IEEE 802.11 network is used typically for covering small areas with a range of up to 200 m. WiMAX is positioned to provide a longer range of up to 20 Km by using advanced antenna technologies. IEEE 802.16 based WiMAX is also designed to use both licensed and unlicensed band.

12.6.2   Comparison with 3G cellular technologies

Current 3G cellular data access technologies allow wide area data access with rates up to 10–15 Mbps. There are two competing technologies in this arena: WCDMA-based high-speed data access (HSDPA) from 3GPP and CDMA2000 based Evolution-Data Only (EVDO) from 3GPP2. Both of the technologies use CDMA-based Physical layer. Revision/evolutions of EVDO with the latest EVDO revision C allows for higher uplink and downlink rates with lower latencies.

The main difference of WiMAX with these technologies is the use of a more efficient OFDM technology at the physical layer. WiMAX also provides a more flexible network design due to being an all-IP technology. The WiMAX standard has much more advanced QoS management provisions. Finally, WiMAX is positioned to support a much higher rate. However, it is not clear if WiMAX can support the same level of fast mobility as handled by the current 3G solutions.

12.6.3   Comparison with LTE and UMB

There are a significant amount of similarities between the LTE/UMB and WiMAX. All of these technologies use OFDM. UMB and WiMAX are completely OFDMA-based, while LTE uses OFDM for downlink. All of these technologies are positioned to be an all-IP network. Both LTE and UMB are also planned to provide up to 100–200 Mbps downlink rate and 50–60 Mbps uplink rate. It remains to be seen, which technology finds wider acceptance in the next generation 4G wireless network. However, WiMAX has the edge over the LTE and UMB by being already standardized and ready for deployment. Recent deployment of WiMAX by Sprint in USA may be an indication that WiMAX might be deployed very quickly for broadband access. The deployment of LTE and UMB is not planned in the near future and may have to wait until the end of 2009.

12.7   SUMMARY

WiMAX technology has been developed with the right set of goals in mind where an all-IP network can be deployed to support various types of application. These include various provisions to meet application-specific QoS requirements and better throughput over a wider range. The two main differentiating technologies used in WiMAX are the OFDMA-based physical layer technologies and the MAC that allows for flexible QoS provisioning. With this set of features, the WiMAX-based network can be easily deployed to carry VoIP traffic, thereby bringing VoIP to last mile mobile users. While this chapter gives a brief introduction to WiMAX, interested readers can refer to [1–3] for more details on the subject.

REFERENCES

1. http://www.wimaxforum.org.
2. Andrews, J., Ghosh, A. and Muhamed, R. Fundamentals of WiMAX: Understanding broadband wireless networking – the definitive guide to WiMAX technology, *Prentice Hall Professional* (February 2007).
3. Vaughan-Nichols, S.J. Achieving wireless broadband with WiMax, *IEEE Computer* 37(6): 10–13 (June 2004).

CHAPTER 13

VOICE OVER WiMAX

For the near future, WiMAX definitely has an edge in terms of a wireless network solution that can support multiple applications and services (Internet, VoIP, Video). The last chapter described the core technologies underlying a WiMAX network that makes WiMAX a Broadband Wireless Access solution supporting high data rate and multiple users. However, an efficient network in terms of throughput and range is not enough when it comes to supporting a diverse set of applications. WiMAX also provides a rich set of QoS management and provisioning features, which, for example, are not present in WLAN. This chapter focuses on providing an in-depth understanding of the QoS related features that exist in WiMAX standards for supporting VoIP. The chapter also provides important design considerations to optimize the deployment of VoIP over WiMAX networks.

13.1 INTRODUCTION

The primary set of considerations in deployment of VoIP over WiMAX are the following:

- service-level differentiation;
- reliability in VoIP packet delivery;
- VoIP capacity.

Since a single WiMAX network is supposed to carry data traffic apart from VoIP, it is required that VoIP packets receive differential treatment. Service-level differential policies

VoIP: Wireless, P2P and New Enterprise Voice over IP   Samrat Ganguly and Sudeept Bhatnagar
© 2008 John Wiley & Sons, Ltd

should ensure that the strict QoS constraint requirement of VoIP is met. At the end, the VoIP packets are transmitted through a wireless medium that is unreliable due to the presence of interference. The unreliability in the wireless medium can degrade the VoIP quality. There should be an efficient way of handling the wireless channel related problems so that both the VoIP QoS is maintained and the VoIP capacity is increased. It is definitely possible to overprovide resources to maintain VoIP quality, but at the cost of VoIP capacity. Techniques to increase VoIP capacity are definitely of interest to any service provider. This chapter provides the latest developments in the standardization efforts and research work to address the above design consideration.

Focusing on the VoIP application, we first provide the reader with a sense of how a WiMAX network deals with a VoIP call. The chapter starts with basic MAC-level operations for placing a VoIP call. Subsequently, the QoS management architecture is introduced to provide a detailed description of how multiple service classes are supported. The QoS provisioning for a given class is tied directly to the underlying scheduling policies. The scheduling policies as proposed in the WiMAX standard are explained, focusing on the specific requirements of VoIP. Next, the chapter describes optimization strategies for reliable VoIP delivery. The chapter also provides a summary view of the performance of VoIP based on reported experimental results.

13.2 VoIP SERVICE DELIVERY OVER WiMAX NETWORK

There are many important protocol components of WiMAX that enable the delivery of VoIP service not just to fixed users but also to the mobile users. We discuss the following parts of the protocol:

- network entry level process;
- inter-BS handoff process;
- power-save features.

This discussion is supposed to provide the reader with the general idea of the operation of WiMAX network for supporting VoIP. For more details, refer to [1].

13.2.1 Network entry process

A subscriber station that wants a VoIP service stream from the base station transmits the ranging request (RNG-REQ) packet that enables the base station to identify the initial ranging, timing and power parameters. Service-flow-parameter requests (bandwidth, frequency, and peak or average rate) are sent next and variable length MSDU indicators are turned on. After receiving a connection request from a subscriber station, the base station transmits a ranging response which provides the initial ranging, timing and power adjustment information to the subscriber station. VoIP-service-flow parameters are agreed on, and a basic connection ID is provided to the subscriber station.

13.2.2 Inter-BS handoff process

The 802.16 standard provides the framework and protocols for supporting the hand-off process. However, the standard does not specify any decision algorithms that relate to which BS to select for handoff. We provide an outline of the framework and protocol.

Each BS broadcasts information about the network topology using the Neighbor Advertisement MAC Management message. This message provides channel information about the neighboring BS. The BS can acquire this information using the backhaul network that connects the BSs. This information helps the Mobile Station (MS) by not having to monitor transmission from all neighboring BSs. Based on the information, the MS may decide to pursue the scanning phase. In this phase, the MS sends requests for scanning to neighboring BSs. Once the request is granted, the MS will attempt to synchronize with the downlink transmission of the neighboring BS and estimate the quality of the PHY channel. During scanning, the serving BS can buffer the data intended for the MS. Following scanning, the MS will associate with the selected neighboring BS after going through the ranging process.

13.2.3   Power-save modes

The majority of the devices that are being used for VoIP services are portable and battery operated. Therefore, a power-saving feature is an integral part of delivering VoIP. WiMAX standards include the following power save mode operations. Power save mode at the MS is achieved through negotiating the period of absence with the BS. WiMAX defines three power-saving classes, based on the manner in which the sleep mode is executed. In Class 1, the sleep mode is exponentially increased from a minimum to a maximum value. Class 1 is applicable for best-effort and non-realtime traffic. Class 2 has a fixed length sleep window and is used for CBR traffic such as VoIP. Class 3 allows for one time fixed width sleep window and is used for MS to perform scanning for handoff etc. where the time window is known. A fourth mode called the idle mode is also available in WiMAX. This mode results in higher power saving. In this mode, the MS can turn off completely, yet receive downlink broadcast traffic. Through a paging process, the MS can be made to wake up when the downlink traffic arrives.

13.3   QoS ARCHITECTURE

A simple version of the QoS architecture for the 802.16 MAC layer is shown in Figure 13.1. The primary differentiation in WiMAX in terms of the QoS provisioning is that the resource allocation is centralized. In the case of PMP (point-to-multipoint) mode, the BS has complete control on allocating radio resources for both downlink and uplink.

As shown in Figure 13.1, both BS and SS maintain connection level queues. In the case of the BS, the incoming packets (MSDU) first enter the downlink queue on a per connection basis. For SS, the incoming packets (MSDU) enter the uplink queue. The job of the scheduler is to serve the respective queues by determining on a frame basis which set of MSDUs should be transmitted next.

From the selected MSDUs, the MPDUs are formed by adding appropriate headers and FEC and pushed to the physical layer. The physical layer creates the OFDMA frames corresponding to per connection MPDU bursts in a TDMA manner and transmits them over the wireless channel. We next describe the process of allocating resources to the downlink and uplink queues.

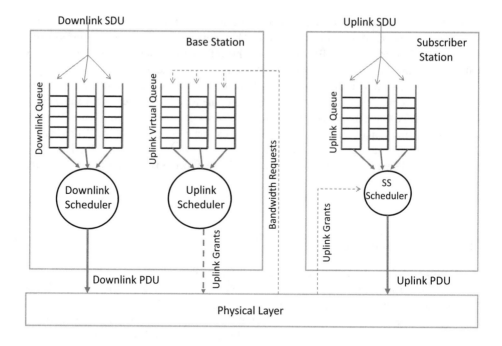

Figure 13.1 WiMAX scheduling architecture.

13.3.1 Serving downlink queues

Serving the downlink queues is simpler. The scheduler works on multiple connections to ensure that per connection QoS constraints on throughput delay are met. For that, the scheduler refers to the connection level specified QoS parameters associated with the corresponding SFID. Based on the criteria of maintaining individual QoS constraints, the scheduler selects which connection to serve next and decides the number of packets to be served for the selected connection. In order to make a correct decision, the scheduler has to be aware of the number of active connections and the number of pending packets in the queue. The BS has all the necessary information available to make the decision. The exact scheduling algorithm is not a part of the standard and is left up to the manufacturers. However, there exists a number of efficient scheduling algorithms that can be used for the downlink such as Weighted Fair Queuing (WFQ) or Weighted Round Robin (WRR). The existing scheduling algorithms can work towards differentiating connections as well as guaranteeing the per connection QoS requirements.

13.3.2 Serving uplink queues

Serving of uplink queues is more complex than serving downlink queues since the BS does both the downlink and uplink resource allocation. The basic structure of the uplink resource allocation process is as follows. The BS holds a virtual uplink queue to estimate the demand for each uplink connection from the SS. The uplink queue refers to the backlogged traffic at the SS. In order to estimate the backlogged traffic, the SS sends bandwidth requests. There are various approaches for sending bandwidth requests, as described in the last chapter.

Based on the bandwidth requested and granted so far, the uplink scheduler estimates the residual backlog for each uplink connection. The job of the uplink scheduler at the BS is to grant uplink frames for each SS by referring to the QoS constraints and the residual backlog. When a SS receives an uplink grant, using the grant information, the scheduler decides to pull a packet from the uplink queue for transmission. It is to be noted that the BS gives the capacity grants for a given SS. It is up to the SS scheduler to decide how to use the grant to serve individual uplink connections.

13.3.3 QoS provisioning

With the above architectural framework, WiMAX can provide a rich set of features for QoS provisioning. In the last chapter, we defined the various sets of QoS parameters that are associated with the service flow corresponding to an active connection. The parameters such as maximum or minimum allowed rate for connections can be easily provisioned and guaranteed by shaping and policing the connection level traffic using the schedulers.

Example scenarios include limiting the number of VoIP flows, or ensuring that the total traffic rate from Internet downloads are below a given threshold, or prioritizing VoIP flow in case of bursty TCP traffic. These above scenarios can be easily supported by applying proper scheduling disciplines both for uplink and downlink connections.

13.4 CALL ADMISSION CONTROL

Two main parameters used in order to enable call admission control at the higher layers are: the Committed Information Rate (CIR) and the Maximum Information Rate (MIR). Both parameters are set for a certain service class and regulate the entire aggregated downlink and uplink flows of a given SS connection. The CIR parameter for a WiMAX system is the bitrate that the network agrees to accept from the user. Flows exceeding the CIR are vulnerable to packet discarding policies at the operator. If the WiMAX network is congested, the BS will typically discard frames on connections exceeding the CIR before frames on connections that are within their CIR. Thus, the CIR provides a crude method for being fair when allocating limited capacity. The second parameter, the MIR, regulates the maximum allowed peak rate of a connection. If the transmission rate exceeds the MIR, all the MAC frames violating the MIR will be discarded automatically; usually, the exact details on how the BS uses discarding policies is proprietary to the manufacturer.

13.5 UPLINK QoS CONTROL

The above section provides a general understanding of the QoS architecture defined for WiMAX MAC. In general the scheduling algorithm is beyond the scope of the WiMAX standard. However, for uplink QoS control, the MAC specification needs to define certain control mechanisms to provide various classes of service. These definitions are important to standardize the interactions among BS and SS in terms of the uplink resource control.

The QoS control mechanism is defined for four service classes. In a broad sense, the control mechanisms define the policies and protocol for BS to grant uplink resources to the SS. The QoS control mechanisms for the four service classes are defined next.

13.5.1   Unsolicited Grant Service (UGS)

The class of service using this policy is required to support applications that have constant bit-rate traffic. The example of such applications include VoIP and streaming media (audio, video). The UGS service is based on having unsolicited grants to the SS. This allows the SS to transmit a frame at periodic intervals without having to request bandwidth. The advantage of this feature is twofold: the packets are guaranteed to be scheduled on a periodic basis independent of the existence of packets from competing traffic; and no delay is incurred in terms of requesting bandwidth and reception of grants. The UGS service allows the service to specify the following QoS parameters: the grant size; interval length between successive grants; and grant jitter tolerated. The last part is important for the BS to ensure that the grants are given to the SS while maintaining a low level of jitter.

13.5.2   Real-time Polling Service (rtPS)

The class of service using this policy is required to support applications that have variable bit-rate (VBR) traffic characteristics. Examples include various streaming media that use VBR encoding. For these applications, UGS will not suffice as the traffic rate is fluctuating. In order to support the above kind of application, the BS polls the SS periodically to provide the SS with the opportunity to request bandwidth grants. The above polling mechanism ensures that the bandwidth grant always arrives at the SS within a bounded time frame. The rtPS service allows the SS to specify the following QoS parameters: polling interval between successive transmission opportunities; and tolerated poll jitter.

13.5.3   Non-real-time Polling Service (nrtPS)

The class of service using this policy is required to support non-real time applications such as large file downloads. These set of applications can have variable packet sizes and follow a bursty bandwidth demand. For nrtPS, the BS does not guarantee the polling interval to the SS. The polling opportunities granted by the BS depends on the current congestion or backlog at different SSs. With this mechanism, the SS typically go through contention or piggybacking to send requests to the BS.

13.5.4   Best effort service

Best effort service is defined to support low priority and low bandwidth elastic traffic such as telnet or HTTP. For this service, BS provides no guarantee in terms of throughput or packet delay. In a sense, this service is defined to allocate the residual bandwidth to the best effort class applications after supporting the above services.

13.6   ENHANCED QoS CONTROL FOR VoIP

It is important to understand the efficiency in transporting voice using the above uplink QoS control mechanisms. The efficiency is in terms of the resource overhead. Typically, depending upon the codec used, the VoIP traffic can be characterized either as constant bit-rate (CBR) or variable bit-rate (VBR). A CBR rate traffic can easily handled by the UGS mechanism. We focus on the supporting VBR voice traffic.

Under normal cases, the voice conversation has a significant amount of silence period and a variable rate encoding is performed to suppress the silence period. An example of such a codec is Enhanced Variable Rate Codec (EVRC) which encodes voice data into four rate configurations $(1, 1/2, 1/4, 1/8)$. The rate configurations $(1, 1/2, 1/4)$ are used during talk spurts while the rate $(1/8)$ is used during silence periods. The corresponding exact bit-rates are (8.55 kb/s, 4 kb/s, 2 kb/s and 0.8 kb/s) respectively. Such variable rate codecs result in higher utilization. Next, we look at the problems in supporting such EVRC traffic using UGS and rtPS service classes.

13.6.1   Supporting voice using UGS

In the case of UGS, the grant size and period are negotiated during the initiation of a voice call. Based on the above, BS assigns fixed size grants periodically to the SS. However, such a scheme is inefficient in supporting EVRC as the grant size remains the same even during the period of silence. This results in a significant decrease in the VoIP capacity. The situation is shown in Figure 13.2 where the fixed grant size is shown by a dotted bar, whereas the actual required size is shown by the shaded bar. Clearly, when the rate is reduced to $1/2$ or $1/4$ and $1/8$, there is a significant wastage of resources.

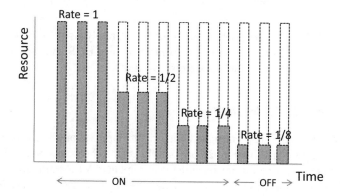

Figure 13.2   Uplink allocation for voice using UGS.

Let us assume a typical voice profile where activity factors for different rates using EVRC is: 29% for rate 1, 4% for rate $1/2$, 7% for rate $1/4$ and 60% for rate $1/8$. This is equivalent to a demand of 40% or a wastage of 60% per voice call.

13.6.2   Supporting VoIP using rtPS

Indeed, for supporting EVRC or variable rate voice traffic, the next best option is to use rtPS, which is actually proposed for such traffic classes. However, rtPS has certain weaknesses in terms of efficiency when applied to carrying EVRC.

In rtPS, the SS has to go through a bandwidth request process where it requests a suitable size grant. This allows exact matching of the variable rate requirement. As can be seen in Figure 13.3, there is no overallocation of resources (the dotted bar matches the shaded bar). However, in order to match the resource allocation, there is an associated MAC overhead in terms of the bandwidth request header. Further, such periodic bandwidth requests also result in extra delay in accessing the resource.

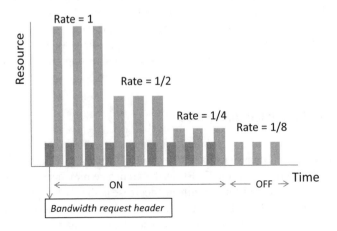

Figure 13.3 Uplink allocation for voice using rtPS.

We note that the periodic property of VoIP is not really used by adopting the rtPS mechanism. A better approach is to have a combination of UGS and rtPS leveraging the positive sides of each approach, which we discuss next.

13.6.3 Enhanced rtPS for VoIP

The Enhanced rtPS (ertPS) [2] is designed primarily to support voice traffic of VBR nature. The ertPS was proposed for IEEE 802.16e recently and has been accepted in the standard. ertPS tries to match the time varying resource requirement of VBR traffic without incurring a periodic MAC overhead as with rtPS. The process of uplink resource allocation in ertPS is described as follows.

First, the voice user informs the BS of his voice status information using Grant Management sub-header in case the size of a voice data packet is decreased. The user requests the bandwidth for sending the voice packets using extended PiggyBack Request (PBR) bits of Grant Management sub-header. In this case, the BS assigns uplink resources according to the requested size, until the voice user requests another size of the bandwidth. The above process is illustrated in Figure 13.4, where the black square denotes the Grant Management Sub-header. As can be noted, the headers appears only when there is a decrease in the rate. Therefore, it can also be noted that the allocation changes at the second interval since the decrease in the rate.

Second, the voice user informs the BS of his voice status information using a bandwidth request header in case the size of a voice data packet is increased. The user requests the bandwidth to send the voice packets using Bandwidth Request (BR) bits of bandwidth request header. In this case, the BS assigns uplink resources according to the requested size and keeps it fixed until the user requests another size of the bandwidth. The BS shall provide the first bandwidth allocation to the next MAC frame after this bandwidth request process.

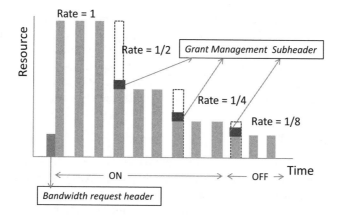

Figure 13.4 Uplink allocation for voice using ertPS.

Therefore, ertPS works like UGS during the time the rate remains the same and does not incur the overhead that is incurred by using rtPS. At the same time, ertPS changes the rate by adding a MAC overhead only during the time of change, thereby increasing the MAC-level efficiency in carrying VoIP.

Performance comparisons of the above three approaches for EVRC codec with 40% activity factor demonstrate the following results. The average assigned uplink resources by using UGS, rtPS and ertPS are 219 bits/frame, 138.79 bits/frame and 113.19 bits/frame, respectively for the single user where the frame refers to the MAC frame.

For details about the performance of the QoS control policies, please refer to the analysis and results in [3,4].

13.7 MAC ENHANCEMENT STRATEGIES

From the last chapter, we learnt that one of the primary functions of the MAC common part sub-layer is to construct a MAC Protocol Data Unit (MPDU). An MPDU is constructed out of fragmenting or aggregation of the MSDU. The MPDU consists of header, payload (MSDU) and FEC. We next explain that the size of the MPDU matters in terms of both the VoIP quality and the VoIP capacity for a WiMAX network [5]. Given the above observation, significant improvement can be gained if the size of the MPDU can be adapted based on network conditions, round-trip time, codec type, etc.

Evidently, the VoIP quality depends upon loss and delay. With both increased loss and delay, the quality deteriorates. Both these metrics need to be considered in terms of the MPDU delivery from the BS to the SS and vice versa to understand the effect on end-to-end VoIP quality.

13.7.1 Packet loss probability

Bad channel conditions result in reception of a corrupted packet at the receiver. A corrupted packet is recovered by applying FEC. If the recovery fails, the packet is lost. Therefore, the packet loss probability depends on the FEC method applied. Generally, the packet loss probability depends on the payload size and the FEC size. Packet loss probability (p) can be expressed as $p = 1 - (1 - b)^{(M+N)}$ where M is the payload size, N is the FEC size and b is the bit error probability. b refers to the current channel condition. Note that increasing M and keeping N fixed increases the packet loss probability. Decreasing N keeping M fixed also increases the packet loss probability. Therefore, we observe that the right choice of the N and M controls the loss probability and consequently affects the VoIP quality.

13.7.2 Packet delay

Though the application of FEC reduces the packet loss probability, the performance can still be further improved if the ARQ mechanism is enabled. ARQ mechanisms are applied if the feedback on channel conditions shows poor state and/or the loss probability is greater than a certain threshold. The maximum number of allowed retransmission for a packet is obtained from its corresponding feedback packet. The main advantage of using the retransmission scheme is to lower the loss impairment at the expense of increased delay. For MAC layer retransmissions, a buffer is maintained for every stream at the transmitting WiMAX BS or SS. This buffer helps in temporarily storing the packets unless and until the packets are restored correctly by the receiver. Clearly, retransmissions add to the end-to-end one-way total delay. If the MPDU size is large due to aggregation, the resulting delay in retransmission will be proportionately higher. Furthermore, retransmissions of large MPDU will also result in increasing the bandwidth used successfully to transmit the MPDU. However, a smaller size MPDU, though better for retransmission, incurs a higher overhead in terms of header size both at the MAC layer and the physical layer. Again, we observe that the size of the MPDU matters to the VoIP quality and capacity.

13.7.3 Dynamic adaptation of MPDU size

Based on the feedback from the receiver, one can design strategies for adapting the MPDU size to enhance VoIP quality and capacity. The crucial feedback includes reported instantaneous channel condition and packet loss due to delay.

If the channel condition is good and feedback shows low packet loss due to delay or packet corruption, the MPDU size can be increased by aggregating more MSDUs and reducing FEC size. The above strategy results in a lower overhead and thereby increases the overall VoIP capacity.

On the other hand, if the channel condition is reported as bad, there might be two options: decrease the MPDU size; and increase the FEC size. If the loss due to delay is high, it is advantageous to increase the FEC. Increasing the FEC will result in a lower number of retransmissions, thereby reducing the delay. However, if loss due to delay is low, increasing retransmission can provide better results.

13.7.4 Performance of dynamic adaptation

The study was conducted with G.729 codec with the R-score quality metric. The simple strategies of dynamically adapting the MPDU size described above can result in 10–20% improvement in the R-score. When increasing the number of VoIP calls, the dynamic adaptation strategy can maintain the target R-score for an extra 25% calls thereby increasing the VoIP capacity by 25%. It was also observed that a maximum of 2–3 retransmissions is appropriate for obtaining the optimal performance.

13.8 COMPARISON WITH COMPETING TECHNOLOGIES

It is important to understand how WiMAX technology performs in comparison to various other wireless technologies in supporting VoIP. Based on performance study reports, it is possible to gain an idea about the number of VoIP calls supported using different technologies such as TD-CDMA, 1xEV-DO, WLAN and WiMAX. The study is based on the resource efficiency of the PHY layer and provides the number of calls supported by 1 MHz bandwidth. For a VoIP application with 20 bytes payload, 40 bytes header and generating packets every 20 ms, the number of calls supported with 1 MHz bandwidth for with different technologies are given in Table 13.1. More about performance comparisons of competing technologies can be found in [6].

Table 13.1 VoIP call capacity comparison.

	WiMAX-OFDMA	TD-CDMA	1xEV-DO	WLAN
Uncompressed	10	6	6	3
Compressed	15	9	10	3

The table shows that the WiMAX OFDMA supports the maximum number of VoIP calls per 1 MHz bandwidth and therefore provides maximum efficiency in terms of physical layer resource allocation. Further, it may noted that by having header compression, the number of supported VoIP calls increases. The efficiency to handle compressed packets is different for different technologies and depends upon the unit of resource allocation at the physical layer. One can observe that WiMAX-OFDMA can support compressed voice with a higher level of efficiency compared to the other technologies.

13.9 SUMMARY

Unlike WLAN, WiMAX MAC is designed to provide QoS to multiple types of application. Recent advancement in the MAC standards such as the eRTPS scheduler for VoIP and other enhancements are ensuring an efficient handling of VoIP in terms of capacity requirement. Analysis based on experimental and simulation results shows that WiMAX can provide a very good performance level in terms of jitter and delay to the VoIP application. Results also demonstrate that performance level can be maintained even under the presence of other contending data traffic. Given the well-designed and flexible framework provided by the WiMAX standard, it is also possible to make further enhancement in the area of scheduling algorithms, admission control mechanisms and ARQ mechanisms for optimizing VoIP delivery.

REFERENCES

1. WiMAX forum, Mobile wimax – a technical overview and performance evaluation (2006).

2. IEEE C802.16e-04/522r3, Extended rtPS for VoIP services (2004).

3. Scalabrino, N., De Pellegrini, F., Chlamtac, I., Ghittino, A. and Pera, S. Performance evaluation of a WiMAX testbed under VoIP traffic, *Proceedings of WiNTECH* (2006).

4. Wang, F. *et al*. IEEE 802.16e system performance: Analysis and simulations, *Proceedings of the IEEE PIMRC* (2005).

5. Sengupta, S., Chatterjee, M. and Ganguly, S. Improving quality of VoIP streams over WiMax, *IEEE Transactions on Computers* (2008).

6. WiMAX forum, Mobile wimax – a comparative analysis (2006).

PART IV

VOIP IN ENTERPRISE NETWORKS

When VoIP emerged in the technology scene, enterprises were skeptical about its ability to provide as reliable, high-quality and feature-rich a service as the traditional PSTN. While the lower deployment and maintenance cost were attractive, the limitations of the VoIP service acted as deterrents. In the last few years, with the VoIP community addressing most of these concerns, enterprise networks have started to deploy VoIP and this trend is expected to continue at a fast pace.

This section deals with two of the most important aspects in an enterprise setting. The two technologies discussed in this section are Private Branch Exchanges (PBX) networks and Network Address Translation (NAT) and firewalls that are present in most of the enterprises. The chapters address how these technologies are used in an enterprise setting.

In the case of PBX, we shall describe the migration towards its IP-based version IP-PBX and how IP-PBXs are becoming viable alternatives for the traditional PBX devices. An important direction this chapter takes is to use Asterisk – an open source IP-PBX, to illustrate these concepts. These topics are covered in Chapter 14.

NAT devices and firewalls are commonplace in today's enterprises. These provide enhanced security and ease the network management task in a typical enterprise connected to the Internet. While PSTN, being a separate network, never had to deal with these devices,

VoIP shares the Internet space with the data traffic and hence has to deal with these devices. We shall gain an understanding of how VoIP tackles the problems posed by these devices in Chapter 15.

CHAPTER 14

PRIVATE BRANCH EXCHANGE (PBX)

Enterprises have been using PBXs for several years to satisfy their own communication needs. Typical PBXs have several features that are useful in enterprise communications. Fundamentally, a PBX is just another telephone exchange that serves a particular enterprise. A PBX that can support VoIP calls is called IP-PBX.

The advancement in technology has made it feasible for a typical home computer to take over the task of a traditional PBX. The functionalities and features of PBX are implemented in software with the increased processor speeds taking care of the Digital Signal Processing (DSP) requirements in a PBX. This is true for IP-PBX devices as well where the software additionally has to deal with VoIP along with traditional voice calls.

This chapter deals with the design of IP-PBX and the issues arising in the IP-based communication setting. We first give an overview of why PBXs are important in the enterprises along with the features they support. We then move on to study the details of the most popular open-source IP-PBX – Asterisk. The chapter concludes with a look at other IP-PBXs currently available.

14.1 PRIVATE BRANCH EXCHANGE (PBX)

A PBX provides several invaluable features that enables efficient communication in the organization as well as significant savings in the telecommunication costs. A PBX is essentially a feature-rich telephone exchange that is owned by an organization. In this section, we detail the basic functions and features provided by a PBX.

14.1.1 Basic PBX functions

It separates the calls which have both endpoints inside the organization itself from those that involve an external endpoint. The idea is to avoid using the telecom service provider's infrastructure for the internal calls. Only the external calls which need to use the infrastructure outside the organization should be routed outside.

This further implies that the number of telephone lines that need to be leased from the telephone company can be reduced. All external telephone lines are connected to the PBX and so too are all the internal telephones. If a call emanates from an internal device to an external number, the PBX will bridge the line from the concerned internal device to an external telephone line, which will be used to connect to the desired external device (through the external devices address, like the phone number). The PBX identifies the call to be directed to an external device when the user dials an external access code (typically, this code is 9). The user dials the external devices number following this external access code without any interruption. On recognizing an external call, the PBX will connect one of the external telephone lines to the internal user device and let the telephone service provider take care of routing the call to the correct device. The other aspect is that an internal device needs to contact another internal number. In this case, the PBX routing logic will detect that the call need not go outside the organization premises. Therefore PBX will connect the two internal lines so that the external infrastructure need not be involved.

Not having direct telephone lines connected to each internal device implies that the PBX has to multiplex the external lines on a demand basis. This means that if there are N_e external lines and N_i internal lines, a PBX allows a scenario where $N_e < N_i$. Now if more than N_e internal devices need to connect to external devices simultaneously, not all devices will get their wish. Some devices will have to be denied connectivity. However, the decision on the number of external lines to request from the telephone company will depend on the expected number of simultaneous external calls that the organization expects and the blocking probability due to insufficient N_e that it deems acceptable. Of course, the benefit of using a N_e significantly lower than N_i shows in the form of a reduced telephone line leasing cost for the organization. A simple example illustrating the typical use of PBX is shown in Figure 14.1.

In the inward direction, when a call from an external device meant for an internal device arises at the PBX, it routes the call to the correct device. This is attained by using the extension number that the external device dials after connecting to the main telephone line that links the outside world to the PBX.

14.1.2 PBX features

Along with the basic task of routing internal and external calls and multiplexing the external lines based on demands from internal devices, PBX provide several features that are of interest to enterprises. The most important of these features are:

- *Call Forwarding*: This feature is used to route calls to another device. The call forwarding logic specifies which device to forward a call to when certain events happen. For example, when the original extension is not answering the call, the call can be forwarded to the voice-mail system.

- *Call Transfer*: An on-going call can be transferred across different extensions connected to the PBX. This happens when we call a customer service department of

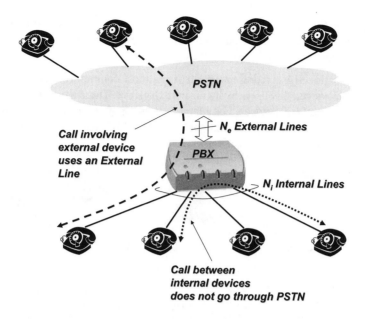

Figure 14.1   The basic usage of PBX.

some organization and the first representative connects us to another to provide some other service.

- *Conference Calls*: The PBX can connect more than two parties into a group communication session called a conference call.

- *Automatic Call Delivery* (ACD): The ACD feature is used to route calls to appropriate handling agents. This could be based on the specialization of the agent, caller priority, etc. ACD may be used in conjunction with Interactive Voice Response (IVR) system which automatically guides users through menus on the telephone asking them to choose among a set of options by pressing different keys on the keypad or speaking certain phrases.

- *Voice Messaging*: This involves recording voice messages for users and then playing it back to them. The message can be for one user who was away when the call arrived or can be left for a group of users as a broadcast message. It can be used in conjunction with the Message Waiting Indicator by which a user knows s/he needs to access the voice message box where one or messages are stored.

- *Call Queue*: In case a call arrives when all the extensions who can answer it are busy, then rather than the call being dropped and the user having to call again, the call can be placed in a queue from which the call will be picked to answer when someone becomes available. This feature can be used in conjunction with the *Music on Hold* feature where the caller whose call that is put in the queue is played back some music (or other information), while s/he waits.

- *Least Cost Routing* (LCR): The calls to the cellular devices can be routed via the PSTN network but will incur additional cost. Alternatively, the call can be routed via a cellular gateway directly making the call cheaper.

The aforementioned list represents perhaps the most commonly used features that a PBX supports. However, this list is by no means exhaustive. There are many more features that are bundled with PBX devices. For a larger sampling of features provided, the reader can refer to the product feature descriptions of various PBX and IP-PBX products from vendors such as Avaya, Cisco, NEC, Fonality, etc.

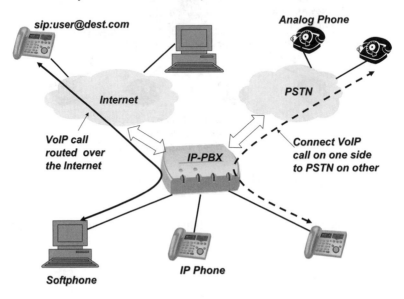

Figure 14.2   The basic usage of an IP-PBX.

14.1.3 IP-PBX

IP-PBX are PBXs that support VoIP calls. The typical IP-PBX also provides a facility to connect a VoIP call to the TDM calls in PSTN. The basic usage of an IP-PBX is shown in Figure 14.2.

From an enterprise's perspective, VoIP offers an excellent communication capability with a fraction of infrastructure cost of the traditional telephony. However, its external customers may very well be connected to the PSTN using analog lines. Thus, while it wants an IP-PBX to handle its internal VoIP network, it also needs the capability to communicate with the PSTN. This naturally means that in order to be of practical use for enterprises, IP-PBX devices need an ability to communicate with the PSTN network.

Like normal PBX, an IP-PBX can also be implemented in hardware or software. In the earlier days when computers were not as powerful, specialized hardware was an integral part of the core of the PBX devices. Faster CPUs have enabled software to take over more and more tasks from the hardware for most of the system.

Perhaps the most popular software-based IP-PBX system is Asterisk. In the next section we will study the design details of Asterisk to gain an insight into the functioning of an IP-PBX.

14.2   CASE STUDY: ASTERISK OPEN-SOURCE IP-PBX

Asterisk is perhaps the fastest growing IP-PBX. It is a software-based open-source system. The advantage of moving from traditional hardware-based PBX solutions to Asterisk comes from its flexibility in configuring the PBX and its ability to run on cheap off-the-shelf hardware. Asterisk is known to provide not just the common features of traditional PBX such as those mentioned above, but also allows users to create new features based on their requirements. This flexibility comes from its implementation being software-based. While the most prominent platform to run Asterisk is Linux, it can be run on Windows, Mac OS X and FreeBSD platforms as well.

Different types of legacy and new telephony devices can be connected to Asterisk. The common analog phones can be directly connected to the host computer running Asterisk by adding FXS/FXO cards for which drivers are available. The ISDN digital phones can be connected on the PRI interface. Asterisk incorporates several protocols allowing several IP-hardphones and softphones to be connected. Due to its software-based implementation, a user can add drivers to provide support for handling any new type of equipment and protocol that come up in future.

The fundamental abstraction used for communication in Asterisk is a *channel*. When there are n endpoints in a communication session, Asterisk opens one channel to each of the n endpoints. This *star* topology abstraction where Asterisk is at the core with all endpoints as spokes, allows Asterisk to perform various sorts of manipulation with the transmitted voice data. In fact, in an extreme case it may be possible that each of the n points uses a different protocol to control its voice session. For example, one endpoint can be communicating using SIP with Asterisk and another using H.323. Still, Asterisk can act as an intermediate server allowing communication between such diverse users. Asterisk supports several communication protocols. These include:

- IAX (Inter-Asterisk Exchange)

- H.323 (ITU Telephony Standard)

- SIP (Session Initiation Protocol)

- MGCP (Media Gateway Control Protocol)

- SCCP (Cisco Skinny Protocol).

A call through Asterisk consists of an incoming connection and an outbound connection. Each call comes in through a *channel driver* that supports one technology, such as SIP, IAX, etc. Each channel driver handles details specific to that channel (e.g. SIP channel driver chan_sip handles all SIP-specific communications). The channel driver creates a private channel for the session and creates a PBX channel for communication. Asterisk then creates a separate channel (using potentially a different driver) to create the outgoing channel for the communication with the other endpoint and then bridges the two channels to complete the connection.

Asterisk supports a bunch of popular codecs allowing users to communicate with diverse technologies. ADPCM, G.711 (A-Law & μ-Law), G.722 G.723.1, G.726, G.729, GSM, iLBC, Linear, LPC-10 and Speex. Like its ability to allow different users to use different protocols, Asterisk also allows different users to communicate despite using different types of codec. This is provided by its transcoding capability. This decoupling of protocols and codecs from the communication sessions allows users with different equipments to be able to communicate through Asterisk.

14.2.1 Software architecture

Asterisk architecture (shown in Figure 14.3) separates the required functionality into two disjoint sets interlinked using APIs. The first set is called *PBX core* which implements the core functionality that determines the internal interconnections of a PBX independent of the specific protocols, codecs, etc. that are used to communicate. The PBX core functionality consists of:

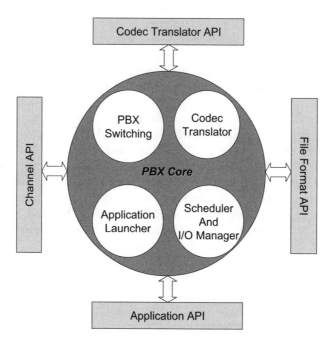

Figure 14.3 The software architecture of Asterisk.

- *PBX Switching:* This connects calls together between various users and automated tasks. The switching is done without knowing the hardware/software interfaces used by different parties involved in the call.

- *Application Launcher:* This launches applications which perform various services such as voicemail, music on hold, and directory listing. Users can write their own applications similar to CGI scripts and configure Asterisk to use those scripts.

- *Codec Translator:* The codec translator functionality is used to convert the encoded voice packets from one format to the other. Each codec module has specified translation capabilities. This facility is exported in terms of encoding and decoding its communication format. The task of translation involves decoding from one format and encoding into the other.

- *Scheduler and I/O Manager:* The Asterisk machine can face different types of load level under different scenarios. This component is responsible for optimizing the performance of the system under different operating conditions.

The core functionality is not concerned with the specifics of the interface or protocols used in the communication session. In order to handle the specific hardware/software that endpoints use and to provide the ability to read/write different formats of data, Asterisk uses *loadable modules*. These modules ensure that the data that they receive/transfer is compliant with the protocol and formats. Due to this design choice, the PBX core portion of Asterisk does not have to worry about the details of individual connections and it can focus on the fundamental functionality.

These loadable modules have the following APIs:

- *Channel API:* The channel API handles the type of connection on which a user connects. This could be a hardware channel (connected device) or a software channel (such as SIP). Each channel abstraction implements the standard specification for that type of channel. For example, chan_sip.c implements the SIP protocol specifications and handles any user that uses SIP to communicate. A channel receives connections, processes the protocol-specific portion, and lets the PBX core functionality provide call-specific PBX services.

- *Application API:* The application API allows various task modules to perform functions such as voicemail, directory listing, call transfer, etc. Therefore, if the processing for a particular call requires a particular application to be executed, this API is used to execute the desired application.

- *Codec Translator API:* This API is used to load the codec modules to support different types of audio encoding and decoding formats. The codec translator uses the encoding/decoding functions associated with each codec module to translate between codecs.

- *File Format API:* This handles the reading and writing of various file formats for the storage of data in the file-system. For example, the voice clips used to create a menu in an IVR can be stored in the gsm format. The file-format API is used to read the gsm files containing these audio clips.

Using these APIs, Asterisk separates the core functions of a PBX server system and the different technologies in the telephony arena. This modular structure allows Asterisk to integrate both traditional PSTN technology and the VoIP technologies. Also, this allows Asterisk to remain future-proof as it requires the creation of independent modules to implement any future technologies without requiring changes to the core functions. For example, if a new protocol is to be added to Asterisk, the only requirement is to create a new channel for that protocol using the channel API. Similarly, in order to create user-specific applications, the Application API can be used to incorporate the new application logic.

14.2.2   Asterisk operation

We now give an overview of how Asterisk functions. The goal of this discussion is to highlight how Asterisk is configured and how a call is processed in the Asterisk software. For the purpose of this description, we will assume that the communication protocol that the user is using to connect to Asterisk is SIP. As we discussed earlier, a user using another protocol merely changes the channel that Asterisk invokes. The use of SIP is merely to serve as an example.

The configuration for the SIP channel is stored in Asterisk code in the sip.conf file. Similarly other channels also have their own configuration files such as iax.conf for IAX protocol and zaptel.conf for the Zaptel hardware used to connect phones to the Asterisk machine. Suppose we have two users, Alice and Bob, who both reside within our organization. Alice and Bob will be using SIP as the base protocol for communication.

As a first step, the SIP channel driver has to know that Alice and Bob are potential users who may use SIP. This is attained by updating the information stored in the sip.conf file that the SIP channel will use. The following shows the addition to the sip.conf file:

```
[alice]
type = user   ; allowed to send calls but can't receive calls
secret = alice_password   ; password for alice
context = internal_call ;the communication context alice can use
host = dynamic   ; requires host to register

[bob] type=friend   ; can both send calls and receive calls
secret=bob_password   ; password for bob to authenticate
context=internal_call   ; the communication context that bob can
use host=static   ; the host need not register
```

The value user for the type field tells the SIP channel driver that Alice is allowed to send only SIP calls. The SIP channel driver will not allow any SIP calls meant for Alice. On the other hand, the friend value corresponding to Bob indicates that he can both receive and send calls. In addition to these two values, a third value peer identifies entities to which Asterisk can send a message. The secret field is self-explanatory. The context field sets the identity of the context where the instructions for this extension are executed. This context is defined in extensions.conf file that contains the dialplan for this extension. Having set host as dynamic implies that the host from which Alice logs in can have different IP addresses and hence it needs to register before starting a session. In contrast, Bob's host has a static address implying that Bob will log in from a fixed IP address.

We note that there are several other options available to configure the SIP channel. We are showing only a small sample of the capabilities here. For different channels the types of options available vary.

The details of the context are stored in another file called extensions.conf. This file contains the *dialplan*. The dialplan describes comprehensively the actions that need to be taken when a call arrives. From the perspective of operation, the dialplan stores the core logic that governs the behavior of Asterisk in order to process a call. In some sense, it is the dialplan that gives Asterisk its power and flexibility. Note that any PBX has some form of dialplan at its core that allows it to perform various call handling tasks. However, Asterisk gains an advantage from being relatively easy to configure and providing a simplistic interface to modify dialplan according to the user requirements.

We now show a possible dialplan for the context internal_call used by Alice and Bob in the example we used above:

```
[internal\_call]
exten => alice, 1, Dial(SIP/alice,5)
exten => alice, 2, Playback(vm-nobodyavail)
```

```
exten => alice, 3, Hangup()
exten => bob, 1, Dial(SIP/bob,5)
exten => bob, 2, Playback(vm-nobodyavail)
exten => bob, 3, Hangup()
```

This dialplan tells Asterisk that when it receives a call for extension Alice, there are three actions it could take (given by the three lines of text for `exten => alice`). The priority of the three actions is given by the second field indicating that the three lines be executed in order. The first line is executed first when a call arrives for Alice. It calls the `Dial` application to call Alice on the SIP channel. It waits for 5 s (the second parameter for the `Dial` application) for Alice to answer. If she does not answer by then, Asterisk executes the next command (indicated by value 2 in the second field that represents the priority of this command for this extension) where it calls the Playback application to play the audio file vm-nobodyavail.gsm that comes with Asterisk package. The audio in this file is 'Nobody is available to take your call at the moment'. We can also have custom sounds played back. After this, Asterisk hangs up the connection. It does a similar thing for all calls received for Bob. When Alice starts her client, she will authenticate using the name Alice and password alice_password. When she sends a SIP_INVITE meant for Bob, Asterisk checks whether she has authenticated and if she is allowed to place calls. If so, it creates another channel to Bob's client and transmits the SIP_INVITE to his client. It acts as a B2BUA for this connection. It also determines whether the two have compatible codecs (for example, if they have different clients, they can have different codecs). If so, it sends a reinvite to Bob to connect directly to Alice so that it can get out of the media-path. If not, it remains in the media path and bridges the two channels to create a communication. This process can be very CPU-intensive.

At the lower level the following happens when an incoming call arrives at Asterisk:

- A call arrives on a channel driver interface (e.g. SIP driver).

- The channel driver creates a PBX channel and starts a pbx thread on the channel.

- Asterisk executes the dial plan for the given context.

- If Asterisk answers the call, it can playback a media (e.g. nobody available as in the previous example).

- If the dialplan requires Asterisk to create an outbound call using the `Dial` application then `Dial` creates an outbound PBX channel and asks one of the channel drivers to create a call.

- When the call is answered, Asterisk bridges the media streams so that the caller on the first channel can speak with the callee on the second channel. In some cases where we have the same technology on both channels and compatible codecs, a native bridge is used. In a native bridge, the channel driver handles forwarding of incoming audio to the outbound stream internally, without sending audio frames through the PBX.

In SIP, there is an 'external native bridge' where Asterisk redirects the endpoint, so audio flows directly between the caller's phone and the callee's phone. Signalling stays in Asterisk in order to be able to provide a proper CDR record for the call. We can create different types of application for PBX by creating the dialplan accordingly. In fact, we can write our own applications to be used in the dialplans. This can be done using the application gateway interface described next.

14.2.3 Application gateway interface

Asterisk allows external application programs to control communication through its Application Gateway Interface (AGI). Instead of exporting an API for programming, Asterisk allows communication with AGI scripts over the standard communication channels STDIN, STDOUT and STDERR. An AGI script receives data from Asterisk on the STDIN channel, it writes data for Asterisk on STDOUT channel, and the AGI script can send any error/debugging messages to be printed on the Asterisk console using the STDERR channel.

The pattern of communication between Asterisk and AGI scripts follows the following sequence:

- Asterisk starts the communication by sending a list of '<agi_variable>:<value>' pairs (describing the environment and context) to the script terminated by a blank line. For example, these variables can be:

```
agi_request: example.agi
agi_language: en
agi_context: local_call
agi_extension: alice
agi_priority: 2
```

- The AGI script takes control of the dialplan and starts sending commands to Asterisk by writing it on STDOUT. Asterisk responds by writing a response on the STDIN for the AGI after executing each command.

- The interaction defined in the previous step continues until the AGI script terminates.

- After completion of the AGI script the control of the dialplan again rests with Asterisk and its execution proceeds normally.

The AGI script can be written in any of the common programming languages such as Perl, Python, C, Java, etc. Asterisk calls an AGI script (say, example.agi) using the application launcher. For example, if a call from Alice in our previous example needs to be processed using example.agi, the extensions.conf entry would be:

```
[local\_call]
exten => alice, 1, Dial(SIP/alice,5)
exten => alice, 2, AGI(example.agi)
exten => alice, 3, Hangup()
```

For example, we can have the following example.agi script written in Perl:

```
#!/usr/bin/perl

use Asterisk::AGI;

$AGI = new Asterisk::AGI;

my \%input = $AGI->ReadParse();

print STDERR "AGI Environment variables and values:\n";
```

```
foreach $i (sort keys \%input) {
    print STDERR "key = $i value = $input{$i}\n";
}
```

```
#Start processing
```

```
print STDERR "Streaming a File"; #Message to STDERR goes to
Asterisk console
```

```
$AGI->stream_file('auth-thankyou'); #print goes to STDOUT. Play
the file "auth-thankyou" to user (Says "Thank You")
```

The above script simply echoes back the AGI environment variables back to the user in the form 'key = AGI_KEY value = CURRENT_VALUE'. It then prints the string 'Streaming a File' on the Asterisk console and plays a file that says 'Thank you' to the user. While the aforementioned task can be accomplished in basic Perl directly, the Asterisk::AGI package provides a nice wrapper API on the basic Perl commands that can be used to interface with Asterisk. For example, the ReadParse function is a wrapper for a piece of code that reads the input from STDIN using which Asterisk passes the AGI environment variables, parses the list of variables, and stores the values corresponding to each variable in an associative array indexed by the environment variable name.

While the above program is very simple, it merely illustrates Asterisk's ability to pass the call control to an AGI script which has well-defined communication interface with the Asterisk system using STDIN, STDERR and STDOUT. The complexity of processing that can be encoded in an AGI script is essentially unlimited. AGI processing can include linkages to databases, reading various files, communicating with other machines for different applications, and so on. We can obtain any information from Asterisk (e.g. DTMF inputs, media stream, etc.) and ask it to execute any command. Similarly, we can use external sources such as LDAP directories and authentication databases to obtain any other desired information. The AGI script uses all these things to implement possibly complex call processing logic. However, one should keep in mind that there should not be much delay in processing since the communication with the user is in real-time and AGI processing adds latency. Hence, writing an AGI script involves a trade-off between the functionality it provides and the latency it introduces due to the corresponding complex processing.

Another aspect that needs to be carefully handled is the use of wild-cards in the dialplan. Simple misconfigurations can have drastic results. One common example is the possible error in the dialplan which can allow anyone to use any valid user-accounts (using long-distance without authorization). If we use, _. as an extension in the extensions.conf file, it will match any extension. Thus, if the corresponding application allows use of long-distance dialing, anyone can use it. Note, however, that Asterisk gives a warning whenever this syntax is used.

14.2.4 System requirements

Having looked at the operation of Asterisk, we now look at the issues that determine the type of system on which Asterisk should run. Asterisk runs on a standard computer instead of being a special device as in the case of the traditional PBX. This mandates that special attention be given to the configuration of the machine that will host Asterisk.

Fundamentally, the technology enabling the usage of Asterisk as a software-based PBX with complex call processing capability is the increase in CPU speed of the modern computers. This enabled the paradigm shift where the traditional media processing tasks that had to be done in hardware to meet the real-time requirements are now doable in software. Asterisk exploited this advantage well and established a presence in the IP-PBX market.

Among other considerations, when selecting the hardware for an Asterisk installation, one must bear in mind this critical question: how powerful must the underlying system be to serve the organization's demand? This is not an easy question to answer, because the manner in which the system is to be used will play a big role in the resources it will consume. There is no such thing as an Asterisk performance-engineering matrix, so the user will have to understand how Asterisk uses the system in order to make intelligent decisions about what kinds of resource will be required. In order to make this decision, we need to consider several factors. These include:

- *Number of Connections:* The maximum number of concurrent connections the system will be expected to support is an obvious determining factor for the machine configuration. Each additional connection increases the workload on the system. This is because Asterisk has to maintain a state for each connection, act as the media bridge for a subset of connections, and maintain communication channels with the endpoints of each connection. Thus, the resource usage increases in direct proportion to the number of simultaneous connections.

- *Conferencing Requirements:* While a simple call has only two parties, conferencing involves multiple parties. At the very least, Asterisk has to mix each individual incoming audio stream into multiple outgoing streams. Mixing multiple audio streams in near-real-time can place an enormous load on the CPU. Further complexity is added if different users are using different codecs whence Asterisk will also have to transcode the audio streams. Thus the choice of system on which to deploy Asterisk is also dependent on whether the provider intends to provide conferencing and if so, how many such conferences can happen in parallel.

- *Echo Cancellation:* When a call involves an endpoint in the PSTN, the speaker can hear the echo of his voice. This is due to the interference among the wires used to provide the traditional telephony lines and the effect of the latency of signal propagation. Echo cancellation may be required on any calls where PSTN interface is involved. Since echo cancellation is a mathematical function, it incurs direct CPU load. The more the echo cancellation requirement, the more the computation load on the CPU. Thus, whether the Asterisk system connects to the PSTN or analog lines also affects the system choice.

- *Call Processing Logic:* Whenever Asterisk has to pass call control to an external program, there is a performance penalty. When we use AGI scripts for complex call processing, then the CPU requirement is high. In fact, if the complexity of the AGI scripts increases, then the per-call computation requirements increase proportionally. Furthermore, the interaction between Asterisk and the AGI script consumes even more resources. The scripts used for processing calls should be designed with performance and efficiency as important goals as the scripts involved also determine the configuration of the required machine.

- *Codecs and Transcoding Requirements:* A codec is a essentially a set of mathematical rules that define how an analog waveform will be digitized. Different codecs trade-off data compression with the quality of signal. Since Asterisk does DSP in software, it has to do more work when a codec has high compression. On the other hand, uncompressed codecs put far less strain on the CPU (but require more network bandwidth). Codec selection must strike a balance between bandwidth and processor usage. Asterisk is capable of transcoding media from one format to another. The decision whether or not to transcode a stream is determined by the type of outgoing channel. If both he incoming and the outgoing channels use the same codec, then there is no transcoding, otherwise Asterisk does it automatically. A major portion of CPU usage of Asterisk comes from its DSP being done in software. Since different codecs have different amounts of processor-intensive DSP requirements (e.g. highly compressed codecs such as G.729 require a great deal of processing and lower compression codecs such as G.711 require less), the number of simultaneous calls that can be supported by Asterisk when transcoding is required depends on the traffic mix:

  1. The fraction of traffic requiring transcoding among all the calls routed through Asterisk.

  2. The fraction of transcoding-requiring traffic that requires a great deal of processing. This depends on the codecs that are in use.

As of version 1.4, Asterisk supports the following codecs: ADPCM, G.711 (a-Law & μ-Law), G.722, G.723.1, G.726, G.729, GSM, iLBC, Speex, Linear and LPC-10. In order to transcode codecs among different channels, Asterisk converts each of them to the linear format and reconverts from the linear encoding to the desired format. Some codecs are computationally intensive and require significant resources. In Table 14.1 we show the relative cost (in milli-seconds) of conversion from one codec format to another on our Asterisk machine (3 GHz machine, 1 GB RAM, RedHat Enterprise Linux, kernel 2.6). We can see that transcoding to iLBC can be very expensive.

- IRQ Latency: Interrupt request (IRQ) latency is basically the delay between the moment a peripheral card (such as a telephone interface card) requests that the CPU stop what it is doing and the moment when the CPU actually responds and is ready to handle the task. Asterisk's peripherals (especially the Zaptel cards) are extremely intolerant of IRQ latency.

14.2.4.1 *Summary*
As we see, there are several factors that can govern the design of an Asterisk system. The exact impact of each of these factors is not exactly quantified. Based on several user experiences, the following observations hold true:

- Transcoding in general is very CPU intensive. Transcoding to certain codecs is much more expensive than others.

- Protocol translation has been observed to be less expensive than codec translation.

- If the machine runs at a load of 5.0 or higher, all VoIP channels suffer.

- Conferencing uses up more resources than simple two-party calls.

Table 14.1   Observed codec conversion time for some codecs on Asterisk on our implementation using a 3 GHz Pentium machine with 1 GB RAM and running RedHat Enterprise Linux.

	GSM	G.711(μ-law)	G726	ADPCM	LPC-10	iLBC
GSM	—	2	3	2	5	19
G.711(μ-law)	3	—	3	2	5	19
G726	4	3	—	3	6	20
ADPCM	3	2	3	—	5	19
LPC-10	4	3	4	3	—	20
iLBC	5	3	5	3	6	—

As a general guideline, since Asterisk is CPU-intensive, the most important component of the host machine is the CPU. A rule of thumb to design the host machine is to use the fastest possible processor and to use a motherboard with fast system bus. Also, it is advisable to run Asterisk on Linux as it is supported officially.

14.2.5   Asterisk as an application server

There are several instances where Asterisk has been used as only an application server with the user-management portion of the setup being delegated to other servers. This type of setup is required in case of the organization having a large user-base. The design choice of using Asterisk as only an application server is known to have increased the scalability of the architecture.

Fundamentally, one can think of Asterisk as a software that has the capability of DSP, audio streaming, and understanding different protocols. Thus, applications can be built using these capabilities while leaving the task of interacting with the users to more specialized systems. For example, SIP Express Router (SER) can be used as the SIP Registrar and Proxy instead of using Asterisk. As such from the SIP architecture perspective, Asterisk is not a SIP proxy as it acts as an end-node itself, making it act like a Back-to-back User Agent (B2BUA). When using Asterisk purely as an Application Server, SER can be used as the front-end and can forward the required calls to Asterisk. SER can use the record-route option to remain in the path of communication. This allows Asterisk to be used only when it is required (to run some special application) and not have to process any extra messages. This partitioning of tasks between different entities makes the system more scalable as each task is delegated to a server that can best handle it.

Asterisk can support several PBX features including voice mail, conferencing, call transfer, etc. A comprehensive list of features is available at the Asterisk website http://www.asterisk.org/features. Each of these tasks along with complex AGI scripts are applications that Asterisk can export as an Application Server. The flexibility that Asterisk possesses to its software-based nature comes in handy in writing innovative applications. With the possible use of Asterisk purely as an application server, the Asterisk machine has to handle fewer messages and hence has a lower load. This margin in load can be used to provide more complex applications rather than having to have small AGI scripts.

The programmer has several tools available from Asterisk: the ability to communicate using different codecs and different protocols, the ability to use media playback (or text playback using the festival utility), and the ability to communicate with different types of device. Therefore, any application processing logic that a user can build is only limited by what a normal computer can do. For example, a simple application could be to playback the e-mail of a user using remote telephony. In this case, the user can authenticate with Asterisk using DTMF signaling, Asterisk can use the LDAP directory to find the user's mail servers, use IMAP to read user's e-mails, and play back the mails using an application such as `festival`. One can envision several scenarios of interactive system that are possible given the features that Asterisk provides.

14.2.6 Desirable features

While we have highlighted the design and working of Asterisk in this section, there are a few aspects that need to be addressed for it to be the universal solution for enterprises. In its current form, Asterisk is lacking in the areas of security and handling fax. Any IP-PBX has to worry about securing the contents of the calls both in the data plane and control plane as they are susceptible to the same type of attack that is used against typical Internet-based applications. Furthermore, fax capability is an integral part of communication that the organizations use over their telephony networks. Thus, any IP-PBX has to support faxing to duplicate the useful functionalities of the traditional telephony. The following issues need to be addressed in Asterisk.

- *Fax:* Sending and receiving faxes is an integral part of the communications capabilities of enterprises. Typical PBX devices have the ability to handle Faxes effectively. Asterisk has the capability to receive faxes. It can be the termination point for a particular number (acting as the fax number globally). It also has the capability automatically to detect fax-tones in the middle of transmissions. It stores the received fax in '.tiff' format which could be e-mailed to appropriate users. Similarly, there is an application called Asterfax that can be used to submit a fax-job to Asterisk via e-mail. The problem with the fax capability of Asterisk is that the quality of fax is not very good. Since fax sends an image across the network, having a bad image quality renders it meaningless. To attain a good quality fax, we have to use a codec which has an encoding rate at least equal to that of the target fax machine. Therefore, we can use a-law and μ-law codecs to send data to a V.29 fax modem (rate 9.6 kbps) but not to a V.34 bis modem (rate 33.6 kbps). Furthermore, the delay and packet drops on the VoIP networks add to the problem as the fax machines do not receive the data at a constant rate and can even miss out on data when the VoIP packets are dropped. This can lead to bad quality of fax at the receiver and hence 'Fax over VoIP' requires special treatment.

  ITU T.38 defines the current standard for fax over IP networks. Just like any VoIP IP Telephony call made over the IP network, fax messages are sent as UDP or TCP/IP packets. The IETF RFC 3362 defines a MIME type called image/t38 for encoding T.38 faxes. T.38 alleviates some of the problems faced by fax machines, especially due to the timing constraints. Overall, using T.38 fax quality is known to increase significantly for most of the fax machines. Asterisk does not have full support for T.38 as of now. An open-source IP-PBX which currently supports the T.38 standard is callweaver which is an off-shoot of Asterisk.

- *Encryption:* With IP-PBX devices working on VoIP calls that transfer voice as data packets over the Internet, the calls are susceptible to all attacks that occur on the Internet. For example, someone snooping on the line from the IP-PBX would have access to all the data from the call and can reconstruct the entire communication. If the attacker hijacks the control plane, the call can be routed to the attacker rather than the correct destination. Concerns exist both in securing the control plane and the data plane traffic in any VoIP setting.

Asterisk does not come with full support for voice encryption (neither SRTP nor TLS). The Inter-Asterisk Exchange Protocol (IAX) that allows communication between two Asterisk systems has some support for encryption. Therefore, the typical method for media path encryption is to use a VPN. There is an effort to add encryption for SIP payloads using SRTP.

14.3 SUMMARY

PBXs are of immense use to organizations as they provide rich communication features while reducing the cost. IP-PBX are the PBX devices with the added capability of handling VoIP calls.

To gain some insight into the features and design choices in an IP=PBX, we delved into the details of Asterisk – perhaps the most popular software-based open-source IP-PBX. The popularity of Asterisk stems from its flexibility in creating complex call processing logics easily and yet maintaining a good degree of scalability.

Asterisk uses a channel abstraction to connect with each endpoint involved in a communications session. Sitting at the core of each communication session, it provides the ability to translate protocols and codecs if different endpoints use different protocols and codecs. However, this feature comes at a cost of increased CPU requirements. The more the requirements for transcoding and extra DSP, the more resources are required for an Asterisk deployment to support a large user-base. As a rule of thumb, it is advisable to have a fast CPU, fast front-side bus, and a Linux-based deployment.

There are several experimental deployments of distributed architectures where Asterisk only serves as an application server and some specialized servers (such as SER) are used as the front-ends. The experience with such architectures is reported to be very scalable.

While certain limitations exist with respect to security and full support for fax, Asterisk represents an important point in the evolution of PBXs where it was shown that IP-PBXs can be efficiently implemented entirely in software.

REFERENCES

1. Meggelen, J., Smith, J. and Madsen, L. Asterisk: The future of telephony, *O'Reilly Media*, (2005).

2. ITU-T Recommendation T.38, Procedures for real-time Group 3 facsimile communication over IP networks (2005).

3. Gomillion, D. and Dempster,B. Building telephony systems with Asterisk, *Packt Publishing* (2005).

CHAPTER 15

NETWORK ADDRESS TRANSLATION (NAT) AND FIREWALL

Network Address Translation (NAT) devices are common components of an enterprise network these days. Initially meant as an answer for the shortage of IPv4 address space, they have grown to play an important role in the areas of network security and management. Fundamentally, this is attained using the seamless indirection they provide between a source and a destination host.

While NAT devices have significant benefits for an enterprise, they introduce new problems in the basic IP communication. The indirection introduced by the devices breaks the direct contact between the end hosts. This violates the design principles that underlie the IP architecture. Hence for NAT devices to be of practical use, its fundamental problems need to be addressed.

An alternate set of devices that create similar problems are firewalls. The purpose of a firewall is to allow only authorized traffic to reach the internal network. As we shall see, from an abstract perspective, the problem that a device behind a NAT or a firewall faces is similar – how does a call initiated by an external device reach it? For this reason, the traversal of both these devices is clubbed together as NAT/Firewall traversal.

This chapter deals with the issues arising in a network bordered by a NAT device. Along with highlighting the type of NAT device and the problems they introduce, we discuss some of the proposed solutions to the problem. While we do not explicitly mention firewalls, the traversal approaches apply to both. The goal of the chapter is to introduce the reader with the recent advances in the understanding of NAT/Firewall traversal.

15.1   INTRODUCTION

NAT is a technique that introduces indirection between the source and destination nodes by systematically changing the source and/or destination IP address fields and potentially the transport layer port numbers of IP packets as they pass through the network boundaries. This is accomplished by one or more NAT devices sitting in the middle of a communication path. The NAT device recognizes a packet belonging to a particular end-to-end session using its source/destination IP address and port number fields, looks up a local table to find the new source/destination IP address and port numbers corresponding to the values contained in the packet, and replaces the original IP address and port fields with these values before forwarding the packet. By ensuring that the address translation process is locally consistent (e.g. one-to-one mapping of the original and the new IP address and port number values), a NAT device introduces a clean separation between the source and the destination nodes.

   NAT was introduced as a mechanism to address the shortage of IPv4 address space. With the explosive increase in the number of hosts connected to the Internet, the 32-bit address-space that forms part of the IPv4 design seemed to be running out. The problem was particularly severe in the non-US countries which had a smaller share of the IPv4 address-space address allocated to them. With the ability to translate the IP address and port numbers in the header field in a packet on-the-fly, the NAT devices isolated the two networks on either side. This allowed the enterprise networks to use *private* IP addresses that were allocated in RFC1918 to create full-fledged private networks. These addresses include those of the type 192.168.x.x and 10.x.x.x, and those in the range 172.16.x.x – 172.31.x.x (where x are values between 0–255). These IP addresses are not uniquely assigned to a single machine across the Internet as is the case with a typical IP address. On the other hand, since multiple copies of these addresses might be assigned to different hosts residing in different private networks, they are not *routable*. The routing protocols used in the Internet ensure that a unicast packet is sent towards its destination IP address. With multiple hosts having the same IP address the unicast will become ambiguous if these addresses are routable. Hence, these addresses are never advertised outside the private network where they are used.

   While the original purpose of NAT was to provide relief in the face of the paucity of IPv4 addresses, their use has gone beyond that cause. In fact, even if the Internet completely migrates to IPv6, there are significant benefits of NAT in terms of security and network management functionality that it is likely to remain deployed in enterprises in the foreseeable future. We shall now elaborate on the detailed operation, types and applications of NAT.

15.2   NAT FUNDAMENTALS

A NAT device (NAT box) sits on the border of an internal network inside which the hosts may have private non-routable IP addresses. Note that it is not a must that the internal network nodes have private addresses. The NAT box has multiple interfaces. In the simplest case, it has one interface connected to the internal network and one interface connected to the external network (with a public routable IP address). When a host in the internal network wishes to communicate with a host on the external network, it sends out a packet with the destination IP address set to that of the corresponding host and the destination port being that on which the remote application is listening for the incoming packets. It sets the source IP address and port to that of its local values. Note that this IP address may be a private address implying that any packets that the remote host sends for it are not going to reach it. The NAT box ensures that despite the internal host having a non-routable address, the packets that the remote host sends in response to its packets are routed correctly to the internal host.

The NAT device sees all the packets coming from and going to a host inside the internal network and tries to identify a set of packets as belonging to a *session*. For example, TCP/UDP sessions can be uniquely identified by the tuple: *(source/destination IP addresses, source/destination port numbers)*. In general, different protocols may have different sets of fields in the packet header that may be used to identify the packets belonging to a particular session.

The basic operation that the NAT box performs is to change the IP address and port values in a packet's header so as to make the external host see the packets coming in from a routable address and to make the internal host see the packets as arriving from the address from which it expects to see the packets. This translation is done based on a per-session basis. In order to achieve the consistent translation of packet field values in a session, the NAT device maintains a *lookup table* which stores the per-session entries for the concerned fields. For example, for the TCP/UDP sessions, this table contains the tuple of source/destination IP address and port numbers. The lookup table is used to store information about active sessions passing through the NAT.

These packets arrive at the NAT box on its internal interface. The NAT box bridges the communication between the two hosts by changing the source IP address to its own publicly routable IP address and possibly the port also to a value that it can uniquely associate with the session. For all packets coming from the external host, it changes the destination IP and port fields to the (possibly local) IP address of the internal device and the port on which the internal application is listening. It acquires these values from its lookup table where the connection information for active sessions is stored. For each incoming packet, the tracking data is used to determine the IP address and port to which this packet should be sent on the internal network. The external host only sees the public IP address of the NAT as the other end of the session. For example, a simple version of NAT is shown in Figure 15.1. Here the application on the internal host is communicating on address 192.168.0.21:2345. When its packets meant for an application on $Host_2$ are received at the NAT, it changes the sourceIP:port fields of the packet to 112.34.48.209:1034 and sends them to the destination host. When NAT receives the packets with destination IP:port as 112.34.48.209:1034 from $Host_2$, it overwrites the destinationIP:port to 192.168.0.21:2345 before sending it to $Host_1$. Fundamentally NAT is meant to do this translation only. However, in practice there are several other things involved in the translation process.

Along with header translation, the NAT device also needs to update the checksum that forms part of the packet header. Since the TCP/UDP checksum includes the content of the header, a change in the IP address and port number values requires recomputation of the checksum. This requires significant computation capability at the NAT device. The problem is exacerbated in presence of packet fragmentation where a large packet carrying a TCP segment is broken into smaller packets when traversing a network with a smaller MTU. In this case, the NAT device has to accumulate all the fragments to retrieve the original segment and then recompute the checksum. This introduces extra delay in the session's path.

In summary, there are two main tasks that a NAT device does:

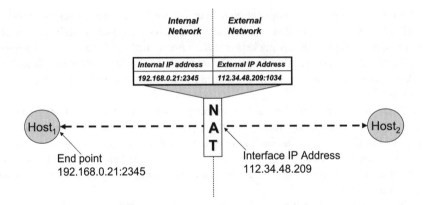

Figure 15.1   The basic operation of a NAT.

- rewriting the packet headers to update the IP address and port numbers to allow internal and external hosts to communicate transparently;

- recompute the checksum for each packet to account for the updated IP address and port numbers.

15.3   APPLICATIONS OF NAT

There are several applications of NAT. While its prime purpose is to temper the effect of the depleting IPv4 address space, the indirection it introduces in the communication can be utilized for several purposes. We list some of the uses of NAT below.

15.3.1   IP address multiplexing

The main task of NATs is to allow enterprises to use very few IP addresses while providing Internet connectivity to a large number of hosts. This feature is meant to provide a relief against fast-depleting IPv4 address space. The hosts inside the enterprise network have private IP addresses. They send packets to external hosts without knowing this. The packets' headers are rewritten by the NAT to reflect its own public address with multiple sessions from internal hosts being multiplexed onto the public IP address by means of different port numbers.

15.3.2   Enhanced security

The use of private addresses by multiple hosts inside the internal network implies that those addresses are meaningless from the perspective of routing from external hosts to them. This ensures that the internal hosts can initiate communication with the external hosts at their discretion (unless the NAT is specifically configured to give a permanent spot in its lookup table for certain services). Second, this limits the ability of any hostile external hosts to perform a port scan on arbitrary machines inside the internal network since they cannot reach them directly anyway.

15.3.3 Load balancing

The hosts inside the internal network may be configured to provide services to the external hosts. In such cases, these hosts are assigned a permanent slot in the NATs lookup table. The NAT device can be used as a load balancer in the internal network. If multiple internal hosts can provide the desired service, then the NAT device can dynamically choose to which server to send the request, based on the current loads of the various servers. This can allow multiple hosts transparently to share the request-processing load.

15.3.4 Failover protection

A more extreme case that NAT can handle is failover protection. This task is similar to the load balancing case where there are multiple servers that can provide the desired service. However, instead of using these devices in parallel and balancing the load across these devices, one or more of the devices are put as a backup to the main system. Thus, in case the main server fails, there are active backups ready. The NAT takes care of routing the request to one of the currently active servers. In case the primary server fails, the NAT device can reset the path to the desired service for incoming packets to reflect the currently active service. It is easy to see that a combination of load balancing and failover protection can also be used.

15.3.5 Advantages

- NAT allows multiplexing of a limited set of public routable IP addresses among a large number of hosts. This serves as an excellent solution to the increasing depletion of IPv4 address space.

- NAT provides a layer of security by preventing external hosts from connecting to arbitrary internal hosts. This is due to the NAT hiding the IP address of the internal machines from the external world. This also prevents external port scans from trying to find vulnerabilities of internal hosts.

- Another use of NATs is in providing load balancing where the NAT device redirects requests for a particular service to one of a multitude of internal servers that can provide the desired service. The public address of all these servers remains the single public address exposed at the NAT.

15.3.6 Drawbacks

- Internal hosts do not have an absolute end-to-end connectivity with an external host. Several applications and protocols are disrupted due to this.

- Special devices are required to support applications and protocols that are designed with end-to-end connectivity in mind.

- In case there are multiple entry/exit points in a network, all incoming/outgoing packets pertaining to a session must pass through the same NAT box.

- Different flavors of NAT limit the ability to have a single solution from helping internal applications connect with the external hosts.

15.4   TYPES OF NAT

The classification of NAT devices can be done in multiple dimensions. We look at the most common nomenclature used to describe the type of NAT device.

15.4.1   Based on type of translation

The NATs can be classified based on what fields of the packet header they translate.

15.4.1.1   Basic NAT   The basic NAT refers to the simplest form of translation. This involves translating only the IP address field of a session's packets. This means that there is a one-to-one correspondence between the internal IP addresses and the publicly routable addresses. Thus, there is no address reuse in this case.

15.4.1.2   Address and port translation   This is the more common form of translation. In this case, both the IP address and the port number fields for a session's packets are translated. This is what allows the original purpose of NAT to be met where a large number of internal hosts share a limited pool of public IP addresses.

The NAT device multiplexes the private IP addresses and port numbers into unique public IP address and port number pairs. In the extreme case, even having one public Internet address, the NAT device can multiplex 65 536 (16 bits for the port number field means 2^{16} unique port numbers) different unique sessions onto this address. For example, if we have a public address of 1.2.3.4, a session from 192.168.1.31:3174 can be mapped to 1.2.3.4:1 and another one from 192.168.1.43:2234 can be mapped to 1.2.3.4:2. This allows the sharing of IP address space which was the prime purpose of NAT.

15.4.2   Based on session binding

Another type of classification of NAT device is based on how they bind the internal addresses to the public addresses.

15.4.2.1   Static   Static NAT uses a strict binding where a particular private IP:port from the private network is always mapped to the same public IP:port. This type of binding is of limited use if applied across the entire private network. There is no multiplexing of addresses as there is a one-to-one binding. Also, the entries in the translation table here are permanent.

15.4.2.2   Dynamic   Dynamic NAT is the other extreme of the static flavor where all private IP:port numbers are mapped to the public IP:port numbers on the fly. This achieves the public address multiplexing gain for which NAT devices were meant. However, this limits the external hosts' ability to initiate communication with *any* internal host. This is essentially implementing a firewall that disallows all incoming connections.

The entries in the translation table are temporary and remain alive only until the time that the session is active.

15.4.2.3   Hybrid   The hybrid implementation essentially has a portion of the translation table assigned statically and the remainder dynamically. Thus it achieves the dual purpose of enhancing security by disallowing arbitrary incoming connections to reach a certain portion of the internal network while allowing other desirable incoming connections to reach the correct hosts. An example of such an incoming connection is an HTTP request meant for the web server of the organization. If we use a purely dynamic NAT, the public address corresponding to the web server (say whose private address is 192.168.1.28:80) may constantly change. Even if the NAT device has only one public IP address, the associated port number may keep on changing. In a hybrid NAT, the public address associated with such internal services are statically bound. Thus, the NAT device can be configured to use a fixed address (say 1.2.3.4:80) to map to the private address of the web server. The sessions from/to other hosts are mapped dynamically.

15.4.3   Based on allowed connections

This classification of NAT is based on which external hosts are allowed to connect to an internal host which has a session active in the NAT. Each session has a corresponding publicly visible IP address and port number in the NAT's translation table. Depending on which applications can use these stored values to communicate with the internal application determines this classification.

15.4.3.1   *Full cone NAT* Full cone NATs are those which have a one-to-one mapping for each public address to a specific internal address and port. In this type of NAT, any external host can send a packet to the particular public address and port on the NAT and it will be routed to the corresponding internal host. This is basically the static form of NAT.

15.4.3.2   *Restricted cone NAT* Restricted cone NAT introduces one level of restriction over the full cone NAT. While all packets from the same internal IP address and port are mapped to the same public IP address and port on the NAT, an external host sending packets to the corresponding IP address and port will not always succeed. The external host's packet will reach the internal host only if the internal host has previously sent out a packet.

15.4.3.3   *Port restricted cone NAT* Port restricted cone NAT adds one more restriction to the restricted cone NAT. Again all packets from the same internal IP address and port are mapped to the same public IP address and port. However, in this case the external host can send a packet to it on a given port only if the internal host has sent a packet to it from that port.

15.4.3.4   *Symmetric NAT* In a symmetric NAT, each session from an internal IP address and port can be mapped to a unique public IP address and port even if the session is to a specific external IP address and port. A session is assigned separate IP address and port irrespective of the destination IP address and port for which the session is meant. On the reverse path, only an external host that received a packet from the internal host can send a packet back to its corresponding IP address and port.

15.4.3.5   *Summary* Each flavor of the NAT based on the type of connections they allow adds more restriction compared to the prior variant. In practice, most NATs are some combination of various types of NAT described above. It is advisable to refer to the NAT behavior specifically rather when they behave as combination of multiple flavors.

15.5   FIREWALL

While it can be used for several applications as discussed above, fundamentally, a NAT performs address translation. Like a NAT, a firewall sits on the border of an internal and an external network. Its purpose is to allow only authorized traffic from the external hosts to gain access to the internal services. It acts as a barrier which filters out unauthorized traffic. For example, it may allow all traffic destined to the organization's web-server's IP address and directed at port 80 (the HTTP port). On the other hand, it can disallow packets originating from the external network directed towards an internal server that is meant for internal organization use only.

In general, most firewalls allow external packets to reach internal hosts in two scenarios. In the first case, they should be specifically configured to allow traffic meant for certain hosts to get through. In the second case, the internal host should have sent out a packet to the external host in the immediate past (often less than a minute). The first case allows the administrator to allow access to services that are meant to be accessed through the external network such as web server. The second allows common internal clients to access services in the external network when they wish. For example, if a user wants to access a webpage of an external site, s/he will be able to see the content only if the firewall lets through the packets that the destination web server sent in response to the client's HTTP request.

Unless the network administrator statically configures the firewall to allow connections to a specific port for each VoIP capable device in the network, any traffic from an external host trying to initiate a VoIP call to the internal device will encounter the firewall. In general, this may not be possible since a VoIP client may reside on any arbitrary host in the internal network, say as a softphone, and the corresponding protocol may be operating over arbitrary ports. Thus, in order to initiate calls from the external device to an internal device, firewall traversal solutions are needed.

While a firewall may implement complicated security policies, from the perspective of VoIP the main problem is that of automatic firewall traversal to allow calls to internal hosts. Based on how a firewall behaves, clearly the only solution to this problem is to utilize the fact that the internal host has to open a connection to the external network and the external host has to use that opened connection to send traffic to the internal host. Hence conceptually, the traversal problem for a firewall is identical to that of a NAT. For this reason, the traversal across NAT and firewall is not usually distinguished as separate and is clubbed together and referred to as Firewall/NAT traversal. Hence, in the next section we shall only refer to NAT traversal, but the concepts apply to firewalls as well.

15.6   NAT TRAVERSAL SOLUTIONS

We have seen that several types of NAT exist in the Internet. While in the simplest form of communication, the internal host can initiate communication with an external host and any form of NAT will allow the corresponding IP:port pairs to communicate, the main problem occurs when any external host wants to communicate with the internal host. The problem worsens when both the communicating hosts are behind NATs. Thus, since neither node has a public address, nor can they route packets to the other. This mandates the need for some components that are publicly reachable to bridge the gap between the two hosts.

Furthermore, applications which use multiple sessions between the endpoints require that all the sessions belonging to the application be initiated from the internal host. VoIP is one such application where there may be more than one session between a pair of end-hosts. For example, these sessions could be a separate control session to govern the specifics of the call and a data session which carries the actual voice packets.

Given the descriptions of the types of NAT, it can be seen that establishing a connection is relatively easy when only one endpoint is behind a NAT. Requiring the internal host to initiate the session to the external host would work irrespective of the type of NAT involved. The major challenge in NAT traversal is to enable communication between two hosts both of which are behind NAT.

Several solutions exist in the literature to enable hosts hidden behind NATs to communicate. There is a fundamental assumption behind all these protocols – whenever an internal host sends out a packet through the NAT, a connection state mapping is established for it and is open as long as the internal host keeps sending traffic to that destination. This is referred to as a *pinhole*. The pinhole is an opening in the NAT through which an external device can communicate with the internal device. Which set of external devices can access the internal device through the pinhole varies depending on the NAT policy. The traversal solutions fundamentally deal with how to get the internal device to open a pinhole and how to get the external host to use that pinhole.

We will look at the general mechanism proposed by a few of these solutions for NAT traversal. Prior to looking at the methods followed in these protocols, we will look at a general method (proposed in RFC3489 [1] along with the STUN protocol described later) by which the internal hosts can figure out the type of NAT they are hidding. This is important in determining the type of traversal solution that may or may not be used in a given setting.

15.6.1   Determining the type of NAT

There are different types of NAT in the Internet. They impose different levels of constraint on the communication. Depending on the types of constraint, the complexity required at the NAT traversal aiding component varies. If the application knows the type of NAT it is running behind, the required complexity can be optimized accordingly. The following discussion details a possible strategy to figure out the type of NAT behind which the internal host lies. This method is proposed along with the STUN protocol in RFC3489 specifically for the UDP flows. However, the method is generic enough as it is based purely on an external server echoing back the IP:port from which it sees a message coming.

To determine the type of NAT, an external server is used to echo some information from the packets it receives from the internal host. The configuration of the system is simple: The internal host lies behind a NAT and the server is in the public Network. Even if the client is behind a cascade of NATs, the server still sees it as operating from behind one NAT. The client knows of the server's address/port through external means. For example, the server address can be hardcoded in the client or it can be retrieved through DNS. The server also has the ability to listen to an *alternate IP address and port*. It will inform the client of the alternate IP:port in its communication. The client sends certain messages to the server. The server has to echo these messages to the client but after performing certain processing as stipulated by the type of the message:

- *Type #1:* On receiving Type #1 request, the server merely stores the source IP:port it sees in the request packet in the response packet and sends it back to that IP:port address (which the device will receive irrespective of whether it is behind NAT or not).

- *Type #2:* The Type #2 request tells the server to send the response from the alternate IP address and different port than the one on which it is operating. Thus, the server receives the request on its main IP:port (and that is the address that may be stored in the NAT) but the IP:port it writes in the response packet's header is its alternate IP:port.

- *Type #3:* The Type #3 request asks the server only to change the port from which it is sending the reply. The server stores its correct IP address in the source IP address field in the reply but fills in the alternate port number in the corresponding field.

Using these three types of message, the internal host can determine the type of NAT it is behind. Observe that the types of NAT (depending on the allowed connections) differ in the binding that they use to route packets for a session. The difference in bindings is based on the IP:port of the source and destination fields and that is precisely what is being modified by the server in the response messages (using its alternate IP address and port). Thus, based on the type of request sent, the internal host would know whether it receives a response on the same opening in the NAT when the IP address or port number changes.

The procedure followed by the client to determine its status is illustrated in Figure 15.2. The client initiates the connection by sending a Type #1 request to the server IP:port. Note that a UDP packet in these requests represents a unidirectional communication. For bidirectional protocols such as TCP there are some subtleties that need to be addressed. If the internal host does not receive any response, that implies that all *UDP communication is blocked* irrespective of who initiates the connection. If it does receive a reply then it is reachable via UDP communication. It next checks whether the IP:port returned in the response (as seen by the server) is its own IP:port. If it is, then there are only two possibilities: either it is not behind any device or it is behind a firewall (and not a NAT). To disambiguate between these two cases, the host will send a Type #2 request to the server where the server sends a reply to it with the alternate IP:port field values. If this message does not reach the internal host, it can conclude that it is behind a firewall that does not allow external hosts to communicate directly with the internal hosts using UDP. On the other hand, if it does receive a reply from the alternate server address and port, it can conclude that it is open to receive any UDP from the external hosts. This can be the case where either it is on the open Internet or it is behind a firewall which allows incoming UDP connections.

The alternative possibility is that the host is behind the NAT. This arises when the internal host sees an IP:port reported by the external server other than its own address (in response to its prior Type #1 request). In this case, it needs to determine the type of NAT it is hidden behind. The host sends out a Type #2 request to the server in this case asking it to send a response from the alternate IP:port. If it does receive a response, it can conclude that it is *behind a full cone NAT* which allows any external IP:port to communicate with the internal host once the internal hosts IP:port mapping is created in the NAT. If it does not receive a response, then we know that the NAT is imposing some type of filtering based on the IP:port it sees in the packet. To test the type of filtering, the internal host sends a Type #1 request to the *alternate IP:port* of the server. If the response it receives contains a different IP:port than the one received in the first instance of request Type #1, then the host can conclude that it is behind a *Symmetric NAT*. This because the only way that the server can see two different IP:port bindings for the requests originating from the same internal IP:port is when the NAT allocates an external IP:port based on both source and destination IP:port. If the host is not behind a symmetric NAT (the responses to both Type #1 requests were identical), then the only possibility is that the NAT may be port restricting or IP restricting.

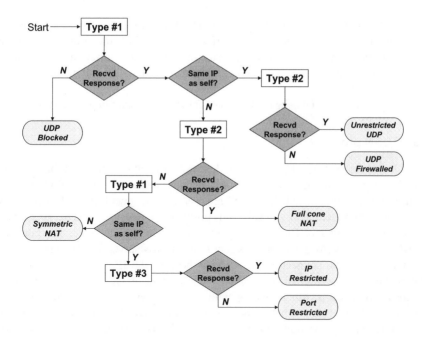

Figure 15.2   The flowchart to determine the type of NAT behind which a host resides.

To test this, the host issues a Type #3 request to which the server responds from its main IP address but with the alternate port. If the response is received by the internal host, then the NAT did not filter the response based on the port. This implies only an *address restricting NAT*. Otherwise the NAT is of *port restricting* type.

15.6.2   STUN protocol

As the name suggests, Simple Traversal of UDP Through NATs (STUN) protocol enables the traversal of UDP messages through NATs. The aforementioned NAT-type discovery method is incorporate in the STUN protocol. The protocol is defined between a STUN server which is running the server component described above and a STUN client which runs on the host (or is part of the application) that needs the traversal. Using STUN, the internal hosts determine their public IP:port and the type of NAT they are behind.

If the internal hosts need to initiate communication with publicly addressable hosts, then they can communicate directly. If they wish to enable incoming communication requests (receive a call), then the host sends a request to the STUN server. If the control component and the data component of the session are decoupled (as they are for VoIP), then the control component needs to help create a NAT binding for the data component too. It does so using a STUN message sent to the STUN server using the IP:port of the data component. Unless the NAT is symmetric, any external host can send data to the internal host using the binding created in the NAT. The NAT bindings do expire after a while if there is no connection. Hence, periodic refresh messages are sent by the internal host to keep the binding alive in the NAT's lookup table.

In summary, STUN enables the opening of the IP:Port for hosts behind NAT by serving as a publicly routable endpoint for incoming messages. It helps keep the port open for any other device to communicate with the host behind a NAT. Since the other hosts contact the internal host directly on the corresponding IP:port on the NAT, the communication fails when the NAT is symmetric. This is because the other host has a different IP:port than the STUN server, resulting in its packets being dropped due to the different binding created in the NAT's table.

For more details on STUN, the reader can refer to the corresponding IETF RFC 3489 [1].

15.6.3 TURN protocol

Traversal Using Relay NAT (TURN) is another protocol that tries to solve the limitations of STUN. The problem with STUN is that it cannot handle symmetric NATs since the public IP:port it sees for a host behind NAT are specific to its own IP:port. TURN involves an additional level of complexity to solve this problem. TURN *relays* the packets between hosts to avoid this problem. Thus, when two hosts wish to communicate, they connect to the TURN server only. All their traffic is sent to the TURN server which relays the traffic to the other end of the connection. This works because the IP:port at the NAT for the connection between the internal host and the TURN server remains the same, irrespective of who the internal host is communicating with. The NAT device only sees the traffic being routed to the TURN server and hence the process works even when the NAT is symmetric. However, this imposes additional load on the TURN server, since it is now in the main communication path of all the calls.

TURN is currently available as an internet draft only [2].

15.6.4 Interactive connectivity establishment

As the above discussion suggests, NAT traversal is an important task in the present architecture. Of the techniques available for NAT traversal, the relay-less techniques such as STUN can allow traversal of UDP traffic across several NAT types and have small overheads. However, if we are faced with symmetric NAT, this simple technique fails. The solution is to use a data relaying technique such as TURN. However, such techniques have significant cost.

In order to choose judiciously which technique to use in a given setting, IETF is in the process of defining the Interactive Connectivity Establishment (ICE) standard. ICE will automate the process of choosing the optimal traversal technique depending on the network environment. As of September 2007, ICE remains an IETF draft undergoing modifications.

15.6.5 Application Layer Gateway (ALG)

While protocols such as TURN and STUN are oblivious of the application semantics, several applications require special handling at the NAT. This may be because of the design choices that were made when the application was initially created and the introduction of NATs changing those semantics. For example, some applications may be sending the IP:port for the destination to reply to in the body of the request message. With a NAT introducing an indirection in the network, the other host sending a reply to the IP:port contained in the request message will fail since that address does not exist at the NAT.

To handle such application-specific traversal problems, the use of ALGs was proposed. The idea behind an ALG is that it understands the application logic and can meaningfully alter the application packets when they traverse through the NAT. The ALG would be co-located with the NAT which offloads each of the application's packets to it. This introduces extra per-packet processing overhead at the NAT.

15.6.6   HTTP tunneling

HTTP tunneling is a method to encapsulate any protocol inside the HTTP protocol. This is useful for an internal device which is behind a very restrictive border device (firewall). In several stringent cases, the network may be configured to disallow communication to almost any ports. However, even in these cases, in practice, access to the HTTP port is typically not blocked. HTTP tunneling takes advantage of this unblocked access to HTTP port.

An internal device in a stringent network wishing to communicate with an external device on a port which its firewall blocks can use HTTP tunneling for the purpose. It will encapsulate its original communication request inside an HTTP packet and send it to an intermediate HTTP server which is located in a relatively open network (which allows HTTP requests to it). The intermediate server will strip the HTTP wrapper and send the actual communication data to the destination external host. When the external host responds to this communication, the mediator server again wraps the response into an HTTP response and sends it back to the internal host. By doing this the external host sees only the communication messages sent in its own protocol and sends to the port where it is listening whereas the firewall sees all the traffic traversing over HTTP.

15.7   NAT TRAVERSAL IN H.323

While most of the aforementioned protocols have been defined in IETF, there has been activity in the ITU forums as well. With respect to NAT traversal in H.323-based systems, three standards have been approved. These are: H.460.17, H.460.18 and H.460.19. Of these H.460.17 and H.460.18 deal with the signaling portion whereas H.460.19 deals with the media transport.

In H.460.17, a TCP connection is opened between the endpoint and the gatekeeper and is used as the transport for all signaling messages. The endpoint which is behind a NAT/Firewall opens a TCP connection to the gatekeeper. The Q.931 signaling messages are sent using this connection and the RAS and H.245 messages are tunneled over it (in both directions). Hence, in this case one TCP connection is opened from the internal host to the gatekeeper and all communication is tunneled over that connection. However, the endpoint needs to know whether this standard is available at the gatekeeper. This can be achieved using a DNS query that resolves the type of service provided by the gatekeeper. If the gatekeeper has TCP service, this protocol works efficiently. However, not all endpoints have the capability to query that of the gatekeeper using DNS.

H.460.18 does not use tunneling. Instead, it opens pinholes from the internal networks on demand. In this case, when the gatekeeper receives a setup request, it signals to the inside node to open up a pinhole in the border device. All communication is now directed through this pinhole by the gatekeeper. This solution does not require the endpoint to have the DNS-querying capability but introduces extra latency and overhead compared to H.460.17.

H.460.19 also works by opening pinholes from inside the network. The recipient of the RTP traffic does this by sending out a stream of keep-alive messages in the form of empty packets sent to the other external IP:port which is the source of the RTP traffic. This keeps open the pinhole for the external traffic to reach the internal host.

15.8 SUMMARY

NAT devices were introduced as a quick fix to the problem of quickly depleting IPv4 address space. However, they have found significant other use, particularly in the fields of security and network management. A NAT device sits on the border of an internal and external network and hides the IP address and ports of the internal hosts from the external hosts. The only IP:port visible to the external hosts are those of the NAT device itself. The NAT device translates the IP:port fields in the packets that traverse through it to help the internal and external host communicate. Different flavors of NATs exist implementing different types of feature and requiring a corresponding degree of complexity. Special protocols such as STUN and TURN are required to help VoIP traverse through a NAT. H.323 has its own set of protocols for Firewall/NAT traversal, namely: H.460.17, H.460.18 and H.460.19.

REFERENCES

1. Rosenberg, J., Weinberger, J., Huitema, C. and Mahy, R. STUN – Simple traversal of User Datagram Protocol (UDP) through Network Address Translators (NATs), *IETF Request for Comments RFC 3489* (2003).

2. Rosenberg, J., Mahy, R. and Huitema, C. Traversal Using Relay NAT (TURN), *IETF Internet Draft* (2005).

PART V

VOIP SERVICE DEPLOYMENT

Until now we have discussed the salient features of different networking technologies and settings and have seen their impact on VoIP deployment. While focusing on these issues, we avoided the specifics of various auxiliary but important issues that did not directly impact our discussion. This section deals with those issues.

The section starts with Chapter 16 which discusses the services and applications that are important from the VoIP perspective. The supporting services such as the use of DNS for lookup of servers responsible for a domain, and ENUM to map an E.164 telephone number to an IP URI are discussed in this chapter. Further, we discuss the applications such as the emergency calling service E911 and fax over IP in this chapter.

Chapter 17 addresses the important issue of security in a VoIP network. It highlights the issues that any VoIP deployment faces and what means we have at our disposal to combat these issues. The mechanisms used in various standards to tackle various security threats are also discussed.

The section concludes with a look into the future with an overview of the IMS standard in Chapter 18. The chapter gives an insight into the design of the emerging IMS architecture and its capabilities that are expected to shape the future of multimedia communication.

VoIP: Wireless, P2P and New Enterprise Voice over IP   Samrat Ganguly and Sudeept Bhatnagar
© 2008 John Wiley & Sons, Ltd

CHAPTER 16

SUPPORTING SERVICES AND APPLICATIONS

In order to deploy VoIP, there are several auxiliary services that it uses. With our focus on concepts rather than details, we have described the protocols and technologies without going into details on the step-by-step procedures they follow. In this chapter we cover some of these services in some detail. The goal is to understand how the VoIP protocols are interconnected with the required services.

We also identify some applications that are typically associated with the telephone network. In order for VoIP to supplant the existing voice infrastructure, these applications need to be supported in VoIP networks as well. We describe the progress that has been made in developing mechanisms to port these applications over the VoIP network.

16.1 DOMAIN NAME SYSTEM (DNS)

We associate the Domain Name System (DNS) as the core Internet protocol that helps in resolving the Fully Qualified Domain Names (FQDN) to IP addresses. For example, when we type `http://www.xyz.com` in our browser, the first step is to identify the IP address of the host on which the HTTP server for the URL `http://www.xyz.com` resides. This is done using a DNS query which is routed across the DNS infrastructure to find out a DNS server which can accurately tell the current IP address associated with that URL. This service is perhaps the most important auxiliary service in the Internet.

However, DNS is not limited to providing a simple FQDN to IP address mapping. This information is stored in the 'A' type record in DNS. There are various other types of record

VoIP: Wireless, P2P and New Enterprise Voice over IP   Samrat Ganguly and Sudeept Bhatnagar
© 2008 John Wiley & Sons, Ltd

by which it can provide various information associated with domains. For example, the 'MX' record helps to deliver e-mail to the proper servers in a domain. When we send an e-mail to abc@xyz.com, the SMTP server does not even know the FQDN of the mail server of xyz.com. In this case, a DNS query is sent for an MX record. This will return the address of the mail server of xyz.com.

A special type of DNS record called the 'SRV' record is used to find the location of services. Protocols can use this type of record to ask the server for a particular type of service. Thus, when Alice places a call for sip:bob@xyz.com, a DNS SRV query is sent out to find out the address of the SIP Proxy server that handles xyz.com. The SRV record containing the details for this proxy could look like:

```
\_sip.\_tcp.xyz.com. 86400 IN SRV 1 30 5060 sippxy.xyz.com.
```

This record indicates that the server for the SIP protocol for domain xyz.com is listening on the port number 5060 over a TCP connection and is called sippxy.xyz.com. The SRV class implies that this record is of service location type. The number 86 400 represents the time-to-live (TTL) for this entry when it is used as a reply. The priority field has a value 1 and the weight field has a value 30. The details of the SRV record are described in the corresponding IETF RFC 2782 [1]. Note that there will also be an 'A' record in the DNS database associating sippxy.xyz.com to its IP address.

The SRV record type provides additional features that can be used for load balancing across multiple servers providing the same service as well as for failover protection. This is done using the priority and weight fields associated with the SRV record. There can be multiple SRV records – one each corresponding to a bunch of servers providing the same service (say, SIP servers). All such SRV records are grouped based on their priority values. Within the SRVs with the same priority value, the servers are used in proportion to their weights. If even one server with a lower priority number is available, the server with a higher priority value is not used by the client. This implies that a multitude of servers in the lower priority class can share the load in proportion to their specified weights and the servers with the higher priority values act as backup servers in case all servers in the lower priority class fail. Consider the following example with three SRV records corresponding to SIP proxy servers:

```
\_sip.\_tcp.xyz.com. 86400 IN SRV 1 30 5060 sippxy1.xyz.com.
\_sip.\_tcp.xyz.com. 86400 IN SRV 1 55 5060 sippxy2.xyz.com.
\_sip.\_tcp.xyz.com. 86400 IN SRV 2 10 5060 sippxy-bkup.xyz.com.
```

Here there are three servers sippxy1, sippxy2 and sippxy-bkup that act as SIP servers for xyz.com. Of these, sippxy1 and sippxy2 both have a priority of 1 and sippxy-bkup has a priority of 2. This implies that sippxy-bkup will only be used if neither sippxy1 nor sippxy2 is available. Among sippxy1 and sippxy2, the probability of choosing them is in proportion to 30:55 (their respective weights). Note that this is a static form of load balancing which does not take into account the actual load of the two servers.

The SRV record is also used in providing information about the servers handling other services in a VoIP network. For example, there could be a SRV record corresponding to the STUN server.

16.2   ENUM

The global telephone numbering system in use in the current telephony system is the ITU-standard E.164. The E.164 standard specifies a maximum of 15-digit phone numbers that define the phone numbers across the world. Along with the length of the numbers, E.164 defines the format of the numbers as well as indicating which digits correspond to the country code and so on.

In the VoIP space, there is a need to locate and call users from the IP network and PSTN. This is enabled by the TElephone NUmber Mapping (ENUM) suite of protocols. ENUM defines the process to represent an E.164 number as an Internet address. This is achieved by ENUM bringing together E.164 and the DNS protocol. The method involves the use of Naming Authority Pointer (NAPTR) records in the DNS [2]. Effectively, it facilitates a seamless connection between the VoIP domain and the traditional telephone system.

In a typical scenario, a user is given an E.164 telephone number that is accessible in the PSTN and a URI name or IP address that is used directly in the Internet portion of VoIP communication. These two identifiers are associated with each other. This binding is stored at a gateway where the simplest strategy is to store all the local E.164 numbers and their corresponding IP addresses in the gateway. While this strategy works, it is not applicable in all scenarios. For example, if the device is using DHCP to obtain an IP address on-the-fly, the static mapping is going to fail.

The technique used to make this binding dynamic yet retrievable quickly is to use DNS for this purpose. The idea is to use the E.164 address as a DNS query and let the DNS server return the correct IP address for the device. The E.164 number is converted into a name to be queried using the digits in the E.164 to form a name. The mapping involves using the complete E.164 number, reversing the digits, and placing a '.' between each digit of the number. Then a string '.e164.arpa' is added to the end of the previously constructed string to obtain the query string. Thus a number +1 234 56 789 will be reversed and mapped to domainname.com which is its parent zone as the name 9.8.7.6.5.4.3.2.1.domainname.com. This zone is delegated to the DNS server of domainname.co. This server will be required to keep the IP address of the corresponding device updated in the DNS database. This information is used to retrieve the corresponding NAPTR record to find out the call preferences of the user. This information is provided in the NAPTR record by the user.

When the caller dials the E.164 number of the receiver, the ENUM gateway will convert it to the associated domain name. Using this information, the corresponding DNS NAPTR record is retrieved to find the receivers call preferences (for example, where the call should be forwarded). Similarly, if the caller is on the Internet rather than PSTN and dials an E.164 number for a VoIP phone, the number would be converted into a URI locally which will be used to fetch the corresponding DNS record.

16.3   NETWORK MONITORING

VoIP is a real-time application with stringent quality requirements. Since the Internet does not provide any mechanism to provide end-to-end QoS guarantees, the VoIP protocols have to adapt to the environment provided by the Internet. To do this, an extensive network monitoring capability is required to keep tabs on the service quality and to be able quickly to isolate the root cause of any problems.

While there are several different methods that can be used for network monitoring, the problem of real-time monitoring and fault isolation is still not solved completely. The challenge in monitoring is to have a non-intrusive system that can monitor the voice traffic while operating at line speed. This gives very little time for the monitoring application to collect meaningful data from each packet that it sees.

Fault isolation is problematic because there are several components involved in the VoIP system. If the voice quality in a deployment becomes bad, monitoring can give an idea of the component which is the bottleneck resulting in poor quality. However, the complex interactions among the components may hide the root cause of the problem. For example, a particular server may appear to be a bottleneck in the system. However, it may be the bottleneck due to a misconfiguration at the load-balancing device which accidently sent a large portion of the traffic to it rather than share it equally among alternate servers. This implies that the monitoring can isolate symptoms but not the root cause of a problem. This is an active area of research.

16.4   DIRECT INWARD DIALING (DID)

On auxiliary service used by enterprises in conjunction with their PBX devices is Direct Inward Dialing (DID). This technique allows businesses to use fewer incoming lines than the number of extensions supported by the PBX. However, each extension inside the organization has a unique directly dialed number to that it can be called from outside the company.

Consider a scenario where the extensions inside the organization have a four-digit internal phone number. However, an external party is given the full ten-digit phone number that can be dialed directly to reach the corresponding phone. The shared external lines are multiplexed among the internal lines on an on-demand basis. The PBX receives the calls and decodes the DTMF tones to identify the called party. Next, it connects the internal phone to an unused external line.

In the VoIP setting too, the number that the service provider gives a VoIP phone is a DID number in a strict sense. The service provider obtains the number from an external source that has a block of numbers and possibly buys these numbers for the use of its customers.

16.5   EMERGENCY CALLING (911)

911 is the emergency calling system in the North American telephone network. Outside the USA a similar facility exits and is often referred to as caller location. When a user dials 911 from any phone, in most cases the call is redirected to a Public Safety Answering Point (PSAP). On receiving an E911 call, a PSAP assimilates the provided location information into its existing database (potentially electronic) and directs the emergency responders to the correct location of the caller. Enhanced 911 (E911) is the feature of 911 which automatically associates the physical location of the caller's telephone number with it. This is usually done using a phone number to address lookup.

At the core of the E911 service is the knowledge of the location of a user. When a PSTN user dials 911 from a phone, it is relatively easy to find the location of the user automatically if the network operator has an updated phone number-to-address directory. An E911 responder has the legal permission to obtain the caller's location using the

Automatic Location Identification function of this database. Along with this database, a comprehensive road and address guide is available for accurate location identification. In the wireline setting, this information makes it quite straightforward to locate an E911 call's origin because in the simplest case, the phone is essentially attached to the location. In the worst case, if the end-user is using a cordless phone, the user is expected to be in the immediate vicinity of the address. This allows the emergency responders immediately to reach the location and address the emergency there.

In a wireless network, the location of a mobile device initiating an E911 call is estimated using the Global Positioning System (GPS) or the radio tracking functionality operated by the service providers. While in the wireline network the location of an E911 caller is in the form of an address, it is a coordinate in case of the wireless networks. In the simplest form, using a cellular base-station to identify the location of a mobile involves finding the location of the base-station itself. This would give the location of the mobile device based on the station to which it is connected. At this level, the device is going to be within a small radius of the base-station's location. More complicated schemes are possible to further localize the position of the device if it is in radio-proximity of multiple base-stations. In case three or more base-stations can sense the radio signals of the device, information such as the observed signal strength measurements can be used to apply standard geometrical triangulation methods to find the location of the device. Existing service providers use several enhancements of these basic techniques to track the location of a mobile device. Once the location of a wireless E911 caller is reliably known, the PSAP to which the call should be redirected is inferred (again based on location). Furthermore, the deduced location of the mobile device is transmitted to the PSAP.

However, providing support for E911 location tracking is not as easy in the case of VoIP. The fundamental problem in this regard is that the VoIP phone is connected to the network over the Internet. In the Internet, a user located in one country can remotely connect to a server located in another country. Moreover, the VoIP call can arrive over other proxy servers. Thus, the problem of associating a location to an E911 call is very challenging.

The service providers are approaching the solution in a phased manner. The first phase of E911 support does not provide location information. It merely routes the call to the 911 administrative phone line. In the second phase, the location tracking functionality would be provided. The effort for this phase is initially directed at providing locations for static VoIP phones. For mobile phones there are still challenges to be addressed. Direct connectivity to PSAPs will be addressed in the last phase.

16.6   FAX

An alternate use of the telephone network is to fax data to another endpoint. A fax machine is connected to the telephone line and has the capability to scan the image of a page to be sent. It dials the number of the destination FAX machine which 'picks up' the call and negotiates the transfer parameters. The sender fax machine sends the image to the destination machine in the form of analog signals. The receiving machine reverses the process and prints out the received image.

For a VoIP network to supplant the PSTN, a similar capability should be available, i.e. a fax machine should be able to send the image data over the VoIP network. While conceptually simple, this does not work well in practice. There are several reasons for this. The simplest reason is the use of a low bit-rate codec for transmission of data. For example, a G.729 codec operating at 8 Kbps cannot transmit signals correctly to a 9.6 Kbps fax modem. However, a G.711 μ-law codec may be able to sustain that rate. For the current V.34bis FAX machines which support 33.6 Kbps transmission rate, the delay in the VoIP channel would be large enough to prevent it from working well (due to echo cancelers not working properly). Additional problems are caused by the silence suppression algorithms used in VoIP as modems need a continuous audio path. Furthermore, the packet-switched and best-effort nature of the Internet network implies that the packets containing signals will arrive at a variable delay introducing significant jitter. In fact, packets can be lost in the Internet resulting in significant damage to the fax. For these reasons, there is a need for a special fax over IP protocols. T.37 and T.38 protocols fulfil this requirement.

The T.37 is a non-realtime fax over IP protocol that defines a method to store and forward delivery of faxes through an IP network. T.37 works well in most settings. However, due to the store and forward nature of the protocol, the two endpoint fax machines cannot negotiate capabilities. The quality of the fax is limited by the store and forward gateway.

T.37 uses the existing protocols such as SMTP and uses them to deliver fax. It defines the procedure for the store and forward gateway to receive a fax, attaches it to an e-mail message and mails it to the remote gateway. The remote gateway delivers it to the destination fax machine.

T.38 provides fax over IP in real-time avoiding the storage latency at the gateways that T.37 incurs. It defines the procedure to contact the remote fax machine over the IP network and sends fax as with the PSTN-based fax machines. It specifies transfer of fax over UDP and TCP. There are guidelines which suggest when to use UDP and when to use TCP. If both the endpoint fax machines are Internet-aware, then TCP-based fax using T.38 will give the best result. However, if one of the machines is not Internet-aware and is connected to the old analog line, UDP is likely to be used. However, UDP packets may be lost. T.38 compensates for this by duplicating the payload of a UDP packet in the next packet. Thus, it requires two consecutive packet losses to lose data. While T.38 improves the quality of fax transmission over the IP network drastically, it still has to cope with the old fax machines that have strict timing constraints to be met. It does a good job of mitigating the problems these machines cause; however, the problems have not been eliminated.

16.7 SUMMARY

The VoIP services need to be integrated into the communication framework and they need to use external auxiliary services for that. The most prominent of these services is the use of DNS to figure out the identity of servers responsible to handle the VoIP protocols at a remote domain. Also, the ENUM service helps in the seamless connection between the PSTN and the IP-based telephony. Network monitoring and fault detection services are needed for the smooth running of a VoIP infrastructure. Features such as DID allow the multiplexing of external lines among a larger number of internal phone extensions behind a PBX while giving an impression that each internal phone has a direct line to it. Applications such as fax are a must for the VoIP infrastructure to become a viable alternative to PSTN. Lastly, VoIP service providers need to develop an ability to locate users accurately to provide the location information to assist in emergency response systems such as 911.

REFERENCES

1. Gulbrandsen, A., Vixie, P. and Esibov, L. A DNS RR for specifying the location of services (DNS SRV), *IETF Request for Comments RFC 2782* (2000).

2. Mealing, M. and Daniel, R. The Naming Authority Pointer (NAPTR) DNS resource record, *IETF Request for Comments RFC 2915* (2000).

CHAPTER 17

SECURITY AND PRIVACY

VoIP has several benefits that stem from the widespread deployment of the underlying IP network. Along with the exceptional quality due to the digital packet-based transmissions of wide-band coded data, the IP-based networks provide a huge user-base and large bandwidth links that can carry innumerable voice calls.

However, the benefits of using the IP networks for transporting voice calls are accompanied by the security risks that any Internet-based application faces. These risks include both those inherent in the Internet and those specific to the VoIP protocols being used. For VoIP to become the ubiquitous voice service, the security measures deployed to tackle these threats must be able to withstand potential attacks. In the traditional PSTN, the number of threats was limited due to the specialized circuit-switched technology that was remote to a casual user. IP networks have a huge user-base which can access services in different parts of the world. This fact in conjunction with the known vulnerabilities of the Internet (and the availability of tools to exploit them) makes the threats to VoIP more real and immediate compared to those faced by PSTN.

In general, it is the VoIP service provider's responsibility to offer a secure service to the user. The service provider must show that the service does not compromise existing security and that the service is on a par in terms of quality and security with the existing alternatives. In fact, not only must the service providers provide a secure service to the user, they must also secure their own networks from outside attacks and service abuse.

The VoIP community has recognized a wide-range of security issues that need to be addressed. These issues may be due to the inherent vulnerabilities of the Internet or to the design flaws of the protocols used such as SIP and H.323. Several solutions have already been proposed and incorporated into various standards and recommendations. However, many issues still remain open and under discussion.

This chapter details some major security issues, their proposed solutions and some open problems in this area. The goal is to inform the user of the possible security threats a VoIP deployment will face. We cover both the generic security issues inherent to the Internet as well as those specific to VoIP and in particular to its protocols.

17.1 SECURITY AND PRIVACY ISSUES

Along with the power to use the Internet for different types of media session, several security problems crop up. Some of the problems are specific to the use of SIP and some are generic. We now detail some of these issues and their solutions in this section. The security issues can be classified broadly into two categories:

- *generic:* the security issues which are prevalent within the Internet and affect other services and protocols as well;

- *VoIP related:* these issues are relevant specifically in the VoIP domain.

We detail some of these issues in this chapter.

17.2 GENERIC ISSUES

The generic security issues are those that are prevalent in the Internet independent of VoIP. Any application that runs over the Internet has to face these issues. Some of the issues are due to the architecture of the Internet and others due to the limitations of the technologies used to access the Internet. Any VoIP service has to deal with these issues. For any service (such as VoIP) to be a viable alternative to an existing solution (PSTN), it must not only be better than the alternative (e.g. in terms of cost, quality of service, etc.), it must be at least as secure and reliable as the existing solutions.

Fundamentally, any device can send a packet to any other device over the Internet owing to its connectivity and routing model. Using this property, a malicious user can actively subvert the VoIP service of a naive user. Furthermore, the ability to intercept communication is easier in the Internet given the type of last hop access technologies such as 802.11-based wireless routers. We discuss below some of the security issues that exist in the current Internet and their implication on VoIP.

17.2.1 Malware

Malware refers to a malicious piece of software like a virus/worm/trojan/ad-ware. Several hosts connected to the Internet are susceptible to attacks because any host can send packets to them. If the packets contain some form of malware and it is installed on the host either knowingly (user opening a malware attachment on an e-mail) or unknowingly (a loophole opened due to a bug in the software on the system), the system is susceptible to abuse.

An infected host can become a source of many problems. For example, the infected host could have been compromised and a back door opened by which an attacker can access the host without any problem. Thus, from afar, the attacker can use the VoIP capability of the host's user. In a simple case, the attacker could use the service to call any numbers s/he wants while the genuine user pays for it. In a more severe case, the attacker could steal the financial information (credit card numbers) from the host and use them illegally. Lastly, an infected host can become the source of infection for other hosts where it can send/install the malicious software on other machines.

17.2.2 Spamming

Spam refers to unsolicited communication that a user receives. At a high level, Internet architecture does not have a provision for having mutual consent of the parties involved before communication could take place. For example, a user can send an e-mail to any e-mail address in the world. The recipient server of the e-mail will receive it using the standard protocols. This e-mail could be unsolicited from the perspective of the receiver. However, the Internet architecture would deliver the e-mail without the notion of consent.

In fact, while there are anti-spamming laws in place, it remains a problem to date. One reason is that the sender of the unsolicited communication may remain untraceable to enforce such a law. The Internet routes packets without worrying about the authenticity of the source of the packets. Fundamentally, the routing algorithm used in the Internet is purely destination IP address-based. Each router looks at the destination IP address of a packet and determines the next hop for that packet. Thus, any source can spoof its own address in the packet header and send any spam to a destination without worrying about being traced back. These days, several ISPs do implement source-address filtering where the packets from sources not in their domain are dropped at the edge itself. However, the spamming problem persists.

With respect to VoIP, the problem has not yet been witnessed in practice. However, unsolicited VoIP calls are likely to become prevalent similarly to spam e-mails or unsolicited calls in the PSTN world. The unsolicited calls existent in the current PSTN could become more severe in the case of VoIP as the caller can easily hide his/her tracks because of the Internet architecture. Furthermore, it is not hard to envision a scenario where a host compromised using some malware can be used to initiate spam calls. A recent trend has anointed such calls as Spam over Internet Telephony (SPIT).

17.2.3 Denial of Service (DOS)

Denial of Service (DoS) is an attack where the adversary overwhelms a server with fake requests so that it becomes overloaded and is unable to serve legitimate requests. DoS is a well-documented issue in the Internet and several incidents have been witnessed. In these incidents, the attacker sends a great many service requests to the server at a rate higher than the server's request processing rate. Soon the server's request queue fills up and all requests start being dropped. Furthermore, the legitimate requests that do get through to the server are delayed by an inordinate amount of time due to the heavy processing load imposed on it by the fake requests.

In the initial attacks, the attacks involve requests with spoofed source IP addresses making it hard to trace the attacker. With the source address-based filtering being deployed at the edge domains, the problem persists primarily due to *botnets* (a large number of compromised hosts) under the control of the attacker. Over a period of time, the attacker uses malicious code to open backdoors into a large number of hosts to create a botnet. In order to launch an attack, all hosts are collectively ordered to send legitimate requests to the target server. This type of attack cannot be avoided using source-based filtering. Universal solutions to such DoS do not exist and it is an open area of research.

With respect to VoIP, the DoS attack is no different. For example, the attacker can cripple a SIP proxy server by sending lots of fake requests, resulting in a genuine user not being able to place a call.

17.2.4   Access technology weakness

We have seen that it is feasible to deploy VoIP over wireless networks such as IEEE 802.11. While this enhances the capability of the VoIP service in the form of user mobility and ease of deployment, it comes with an additional cost. Wireless technologies transmit signals over the air. Thus, not only is the intended receiver of a packet able to listen to it, but also everyone else in the vicinity.

From the perspective of VoIP, this implies that a VoIP call can be overheard by an eavesdropper who is in the vicinity of the source or destination. This results in a significant breach of privacy. While such technologies provide standardized encryption methods, they are not always used in practice. This is unlikely when the end host is in an enterprise setting where the encryption is a must, but very likely when the end user is at home and has deployed an off-the-shelf wireless router for in-home mobility. Furthermore, exploits for such technologies appear, regularly making them quickly obsolete. For example, the Wireless Equivalent Privacy (WEP) standard for encryption used in several wireless LANs has recently been shown to be vulnerable. It is possible to decrypt the WEP-encrypted communication fairly easily by accumulating a small number of encrypted packets.

These vulnerabilities are exacerbated by the fact that the deployment of such technologies is done by inexperienced home users. An expert is likely to configure the devices in a manner so as to limit the possibility of the damage. This was true in the PSTN where the entire deployment (including the wire to the telephone socket in ones house) was done by the service provider. With wireless access technologies, while the experts deploy the lines to the home, the wireless access point is typically deployed by the end-user.

17.2.5   Improper implementation

Many security problems arise due to incorrect implementations of protocols or to software bugs. Perhaps the most common exploit that opens a system to compromise is buffer overflow. If the implementation of a protocol or feature has improper implementation, it may not check the boundary limits for the inputs. For example, if a function expects an array with 10 integers as input and the user passes it an array with 11 values, the 11th value could overwrite the return address from the function, allowing the code to jump to an arbitrary location and allowing execution of an arbitrary command. In the worst case, the attacker could use this to open a shell with root privileges and have the machine compromised.

17.3   VoIP-RELATED ISSUES

While the aforementioned security and privacy issues are generic to all applications using the Internet, there are issues that specifically relate to VoIP itself. These issues are relevant irrespective of the technology used to provide the VoIP service. A subset of these issues is described below. For further details, an excellent source is the VoIP Security and Privacy Threat Taxonomy [1].

17.3.1 Misrepresentation

Misrepresentation implies an act of masquerading as someone else. This includes impersonating another user or another device. The biggest impact of misrepresentation on VoIP is in the form of *Toll Fraud*. Toll fraud refers to an unauthorized user accessing a valid and legal VoIP network. After gaining access to the network, the attacker can make arbitrary calls without having to pay for them. The attacker gains access to the network by misrepresenting his/her identity as another user or as another device by spoofing its address.

As an example, consider the case of impersonating another user using SIP. A malicious user can impersonate any other user since the identity of the caller is simply a value in the From field in the INVITE message. Thus, it is easy for any user to send an INVITE message by using someone else's identity. For example, consider the following INVITE message send by Alice <Alice@domain1.com>:

```
INVITE sip:Bob@domain2.com SIP/2.0
From:<Charlie>Charlie@domain3.com To: <Bob> Bob@domain2.com
```

On receiving the above INVITE message, Bob has no way of knowing whether the message is from Charlie or from Alice. This attack can be used to appear as a third-party but to get the callee to communicate with the attacker (using Via and Contact fields). For example, the following request appears to Bob to come from Charlie, but when he responds, the message goes to Alice via proxy.domain1.com:

```
INVITE sip:Bob@domain2.com SIP/2.0 Via: SIP/2.0/UDP domain1.com
From: <Charlie> Charlie@domain3.com To: <Bob> Bob@domain2.com
Contact: <Alice> Alice@domain1.com
```

A smart adversary can further hide the tracks by spoofing the address of the domain to a greater level. A variant of this attack involves spoofing the BYE message to terminate the call in between. The attacker can send the BYE message to the communicating parties with the message containing the appropriate fields that s/he can eavesdrop and find in the INVITE message. Furthermore, a similar risk is posed by an adversary impersonating a SIP proxy server which could redirect an INVITE to an insecure location.

17.3.2 Service theft

While misrepresentation is used to act as someone else, another form of fraud can occur when a malicious user does not misrepresent any information but bypasses the billing to which the service provider is genuinely entitled. In the simplest form, this may involve modifying the billing information, and in an extreme case deleting it altogether.

17.3.3 Eavesdropping

Eavesdropping refers to an attacker being able to intercept a call in a passive manner. As discussed earlier, this is possible because of access technologies such as 802.11 when the voice packets are sent unencrypted or when the attacker inserts a tap in the wireline link used to connect an endpoint. The latter is similar to the PSTN network where an attacker can tap into a telephone line.

In general, whenever by any means an attacker is able to monitor the signaling and/or the media transport of an endpoint, s/he can use it for various purposes. In one case the information could be used to extract the information about call patterns and in the extreme case it could be listening to the entire conversation.

17.3.4   Call altering

Call altering takes the next step from eavesdropping. While in eavesdropping the attacker is passively listening to the signals, here the attacker can actively introduce traffic in an ongoing conversation. In one form, alteration includes 'Call black holing' where the attacker does not allow certain information to pass, resulting in the call being dropped. Also, the attacker can reroute the call through unauthorized gateways.

17.3.5   Call hijacking

Hijacking implies that some portion of the VoIP service is taken over by the attacker. Call hijacking refers to an attacker inserting a spoofed response to a call initiation request sent by a user. The attacker sends a response redirecting the user's communication to (or through) a rogue server from where the attacker can intercept the call.

Registration Hijacking is one form of hijacking which involves fooling the location registrar to map a user-ID to a separate location. For example, in SIP, this can be done by binding a false address to a URI and sending a REGISTER message to the registrar. This can occur when an adversary captures the original REGISTER message, modifies its Contact field to point to a different location and resends the message to the registrar. This would not allow the genuine end-user to send/receive any calls and all calls meant for the user would be redirected to the attacker's device.

Another form of hijacking is termed *session hijacking*. Here, the attacker replaces media packets while the conversation is ongoing. This type of session hijacking is possible because the media packets are routed over the Internet. That allows a clever adversary to remove the original media packets from the stream and substitute it with other packets. In the absence of countermeasures, the end-users have no way of identifying whether the packets come from the other party or from an adversary in the middle.

17.3.6   Privacy

In certain dialogues, there may be a need to withhold the identity of a person and other personal information from the other party. While such a user can use a dummy value in the From field, the information regarding the proxy from which the message came itself might be enough to identify the person. Furthermore, the issues of providing privacy and avoiding impersonation can lead to conflicting solutions.

17.4   SOLUTIONS

The above problems have been well-documented and some solutions have been proposed. We give a brief overview of some of these solutions and their applicability.

17.4.1   Authentication

One of the key requirements in the aforementioned security problems is the need to authenticate any communication. With respect to VoIP, all parties involved in the communication should authenticate each other. Thus, authentication is required not just for the end-users but an end-user may require a server to authenticate itself before sending it any call set-up information. Both the registration hijacking and impersonation attacks can be solved using a proper authentication mechanism. Broadly speaking, authentication is required during the following processes:

- Whenever an endpoint registers with the authority, the user should be authenticated. In SIP the authority would be the registrar and in H.323 the gatekeeper.

- Whenever a new session is set-up, the involved parties need to be authenticated.

- Whenever the parameters of a session are modified mid-way through the session, authentication should be performed.

- Lastly, whenever the session is to be terminated, authentication is required.

There are several possible methods of authentication. The common method of authentication that SIP provides is called *Digest Authentication* which is similar to that used by HTTP. The digest essentially verifies that both parties know a shared secret (password). When a client connects to a server (for registering or to start a new session), the server responds with an Unauthorized message. The specific type of message differs based on whether the server is a proxy server, redirect server or a registrar. The server's response contains a challenge which comprises a randomly generated value (nonce) along with other parameters such as the algorithm to be used for hashing (default is MD5), the realm of authentication, etc. The client computes a hash of a combination of a set of fields including its username, password and the nonce. This hash is sent to the server, which can verify that that client is indeed the person who s/he claims to be. Note that the use of a nonce is required to avoid replay attacks so that each time the client is asked for authentication, the hashed signature is different.

A similar method is used for authentication in H.323 for signaling messages also. The shared secret between the two parties is a password, preferably encoded as a hash using a cryptographically strong one-way hash function such as SHA1. To prevent replay attacks a 4-byte random value is used (in practice the random value is an incrementing counter) in conjunction with the time-stamp as the perturbation value. The function used for computation of the hash is HMAC-SHA1-96 where the leftmost 96 bits of the SHA1 hash of a bit string is used as the hash value. There is an authentication-only mode where the method only allows for authentication of identity and cannot be used for message integrity check. This is of use when the signaling messages traverse a NAT (and where the packet's IP:port fields would be changed, violating the message integrity).

17.4.2   Message integrity

Authentication ensures the identities of entities involved in the message transfer process but the contents of the messages themselves are not verified. Verifying the integrity of a message is the process of ensuring that a message that was received in the session was exactly what was sent. Thus, message integrity is aimed at eliminating the tampering problem. Broadly speaking, it is an authentication for the message body.

The SIP message body is of Multipurpose Internet Mail Extensions (MIME) type and can be secured using the standard mechanisms used in Secure/MIME (S/MIME). S/MIME provides a consistent way to send and receive secure MIME data. It uses digital signatures for authentication and message integrity, and encryption for data confidentiality. SIP allows the use of S/MIME to encode its body and uses encapsulation to provide authentication of its headers. It allows the entire SIP message to be encapsulated into S/MIME and generates a signature while sending it with the original SIP header. The receiving entity can then verify the SIP header's integrity by comparing the 'outer' message with the inner one. Although S/MIME can provide end-to-end authentication, integrity checking and confidentiality, it requires a public key infrastructure. Furthermore, it can result in large messages.

In H.323 the message integrity check is also done using HMAC-SHA1-96 hashing. The requirement is to hash all the fields of any signaling message. The hashing is done over the ASN.1 encoded message. Thus, if even one bit of the message is changed, its HMAC-SHA1-96 hash is very likely to differ, indicating message tampering. To date, there is no known method which can generate an alternate message for a given SHA1 hash code. More details are provided in the ITU standards document [2].

17.4.3 Signaling message encryption

Although the authentication process takes place using a secure hash, the other fields in the header are publicly visible. This is human-readable in case of SIP where the messages are in clear-text and easily decoded in case of H.323 which uses a binary representation of ASN.1 encoding. Thus, there is a need to hide the content of these fields. Both H.323 and SIP have mechanisms to encrypt the content the messages carry.

SIP is capable of using any security mechanism that the network is willing to provide. These security mechanisms are used for hop-by-hop and end-to-end encryption of certain sensitive header fields and the message body. Examples of such mechanisms include Transport Layer Security (TLS) and IP Security (IPSec). Of these, TLS works only with TCP whereas IPsec is useful for UDP as well. Similar to an https URL in HTTP, SIP also provides a secure URI, called a SIPS URI. An example of such a URI is sips:foo@bar.com. SIP mandates that a call made to a SIPS URI is guaranteed to be secure and encrypted transport (TLS) is used to carry all SIP messages from the caller to the domain of the callee. From there, the request is sent securely to the callee, but with security mechanisms that depend on the policy of the domain of the callee.

In H.323 as well, there is support for TLS and IPSec for encryption of the signaling.

17.4.4 Data encryption

The actual data transfer between endpoints occurs on a separate path than that for signaling. Some form of encryption is required while the messages are being transported so that an eavesdropper does not understand the communication. A recent proposal for real-time transport addresses this issue. The Secure Real-time Protocol (SRTP) is designed for secure communication of real-time applications. SRTP protects both RTP and RTCP packets and guarantees the confidentiality and integrity of each packet. SRTP is designed to have low computational cost (which is a must for real-time communication) and low bandwidth overhead. Furthermore, SRTP is independent of the underlying cryptographic mechanism and hence is amenable to any future upgrades of cryptographic techniques. The default cipher that SRTP uses is Advance Encryption Standard (AES).

H.323 provides support for four different encryption algorithms for voice encryption: AES-128, RC2-compatible, DES or triple-DES. Depending on the desired degree of confidentiality, RTP can be encoded using any of these methods. More details are available in the corresponding ITU standard document [3].

There is a need for a-priori agreement on the encryption keys for data transfer. Multimedia Internet KEYing (MIKEY) has been designed specifically for key management in group multimedia settings. MIKEY is designed to be a low latency protocol suitable for real-time sessions. H.323 supports the use of MIKEY for SRTP [4]. Furthermore, H.323 also requires support for the Diffie-Hellman key agreement protocol.

17.4.5 Privacy

To ensure that users determine their own privacy levels, SIP allows users to hide their identity. However, two complementary principles have guided the design of this privacy mechanism: Users are empowered to hide their identity and related personal information when they issue requests, but intermediaries and designated recipients of requests are entitled to reject requests whose originator cannot be identified if they wish to do so. The question of how much privacy is good enough is still being debated. Some of the solutions are not applicable across the board because SIP has to operate in a wide range of settings. For example, SIP has to account for the users sitting behind a NAT/firewall. In such a case, there has to be an entity such as an Application-Level Gateway (ALG) that understands the SIP protocol and opens/closes appropriate ports inside the firewall to enable communication. This in turn means that the SIP headers have to be in clear-text for firewall traversal (at least some portions of them) and use of encryption mechanisms for hiding SIP headers may be limited.

17.5 RECOMMENDATIONS

Securing the VoIP infrastructure is a continuous process. If a VoIP network is secured at one time, new vulnerabilities may surface in future, either due to new exploits being developed or to change in configurations that may unintentionally have been done. Thus, there are several guidelines that a VoIP network administrator needs to follow. A few of them are as follows:

- Frequently scan the machines in the network for any evidence of malware and if found, quarantine the machine until the malware is cleaned.

- If there are VoIP clients connected to the network using 802.11 wireless access points, ensure that all communication is encrypted using a strong encryption such as WPA2.

- Update the software of all the VoIP components with released patches regularly. These components include not just the softphones but also the server software and the software running on the components such as IP-PBX.

- Properly configure any remote administration capability of any network component. All such remote administration should be avoided if possible. If it is necessary to administer a device remotely, ensure that a strong authentication mechanism and encryption algorithm is used for all communication.

- In general, encrypt all communication with strong encryption algorithms. The computation cost of using a strong encryption and decryption algorithm for each message is minuscule compared to the security it gains.

- Use VLANs generously to separate the VoIP and data traffic in the network. Along with the QoS benefits that the VoIP traffic gains, it helps in enhancing the security of the VoIP network as the computers that are part of the data VLAN do not see the VoIP portion.

- Periodically perform a complete security audit of the entire network and verify that the security policy is properly enforced across the entire network.

17.6 SUMMARY

Security and privacy are important parts of any voice service deployment. VoIP faces additional challenges compared to PSTN as it is deployed over an IP network (Internet) which itself has several security problems. Malicious users can hijack the connections, insert or delete messages, and eavesdrop on the communication in the basic form of communication. Standards have provided methods for authenticating the users, assuring message integrity and encrypting the messages. The use of these techniques is a must to ensure the security of VoIP.

REFERENCES

1. Voice over IP Security Alliance (VOIPSA), VoIP security and privacy threat taxonomy version 1.0, *http://www.voipsa.org/Activities/VOIPSA_Threat_Taxonomy_0.1.pdf* (2005).

2. ITU-T Telecommunication Standardization Sector of ITU, H.323 security: Baseline security profile, *ITU-T Recommendation H.235.1* (2005).

3. ITU-T Telecommunication Standardization Sector of ITU, H.323 security: Voice encryption profile with native H.235/H.245 key management, *ITU-T Recommendation H.235.6* (2005).

4. ITU-T Telecommunication Standardization Sector of ITU, H.323 security: Usage of the MIKEY key management protocol for the Secure Real Time Transport Protocol (SRTP) within H.235, *ITU-T Recommendation H.235.7* (2005).

CHAPTER 18

IP MULTIMEDIA SUBSYSTEM (IMS)

IP Multimedia Subsystem (IMS) is the next-generation network architecture that will facilitate a whole new way of delivering network services to users. IMS was conceived and designed for supporting multimedia services in the cellular world under the 3GPP standard bodies. However, the design principles and motivations behind IMS transcend cellular networks and apply to any IP network. In recent times, the push towards IMS has also been motivated by the fixed mobile convergence with the objective of delivering the service to users regardless of their access network (cellular or fixed). Furthermore, by being a complete IP-based technology, IMS is going to drive the universal adoption of VoIP across network boundaries. IMS is still undergoing evolution in terms of architectural framework and functionalities, and is not yet completely ready for deployment. At the same time, it is important to understand the motivation, basic concepts and design principles that underly the current IMS architecture and that drive the future evolution of IMS.

18.1 INTRODUCTION

IMS was first conceived and introduced under the 3rd Generation Partnership Project (3GPP) [1] early in 2003. The goal of IMS in the context of cellular networks was to create an architecture that streamlines the development, deployment and delivery of new services in the mobile networks. At a later stage, the original IMS architecture was extended by the Telecommunications and Internet converged Services and Protocols for Advanced Networking (TISPAN) standardization body into applying the architecture for

VoIP: Wireless, P2P and New Enterprise Voice over IP   Samrat Ganguly and Sudeept Bhatnagar
© 2008 John Wiley & Sons, Ltd

fixed networks and other WANs. In a sense, the 3GPP focused on the wireless operators, while TISPAN focused on the wireline operators resulting in the fixed mobile convergence (FMC). In this chapter, we describe the IMS architecture that is standardized by the TISPAN [2].

IMS is a functional architecture for service delivery (Voice, Video, Data) based on IP. IMS is based on using different protocols that are standardized by the IETF. The basic goal of IMS is to create an architecture that separates the services from the data transport. For example, data is transported over various NATs to the user including Cellular network, WiFi, DSL, fixed T1 lines, etc. However, using different access technologies to transport data, ultimately the same services are being provided to the users. These services include voice, video, data, etc. Even though it makes sense that the service layer should be common to all the access networks, it is not so in current deployments.

Operators of each access network have their own service layer. Due to this access network division, a user cannot move from one access network to another while availing him/herself of the same service. For example, it is not possible for a user to migrate his/her ongoing voice call through the cellular network to the WiFi network available in his/her office or home. Facilitating the above feature requires the network dynamically to switch the ongoing VoIP call mid-way from one access network to another. The goal of IMS is to enable the service continuity features where the user can continue receiving the same service regardless of the access technology being used for data transport. Thus, IMS provides a converged view of the access networks to both the users and the service operators. No doubt, IMS is providing a natural way to evolve the next generation network that would benefit service providers and users alike.

IMS addresses the architectural and protocol challenges in order to achieve the above goal. What IMS defines is a new functional architecture leveraging many existing protocols such as SIP for delivering service in a network-agnostic manner. This chapter tries to provide the reader with a simple understanding of the underlying IMS concepts and its benefits. The chapter refrains from providing the detailed working of each protocol and encourages readers to follow the standardization work [2] or [3–5].

18.2 ARCHITECTURE DESIGN GOALS

The basic goal of the IMS architecture is to streamline the management of service deployment, delivery and charging. This design goal results in creating an overlay service delivery network on top of a packet-switched infrastructure. The goal also requires separation of control (how the network is managed and used) from the transport (how the data is transported to the user). The basic goals in turn translate into meeting various important design goals, discussed next.

As a part of the main design philosophy, the IMS architecture is network access agnostic. This enables the IMS architecture to be deployed over multiple access networks as a single horizontal layer. The interoperability requirement should enable IMS to handle different types of application, network equipment and device. This requirement results in IMS adopting open protocols designed for IP networks. An important goal to enable network convergence is the roaming capability. This goal requires IMS to support user-level roaming across different network boundaries defined either by the administrative entities (carriers) or the access technology used. The IMS, initially designed for multimedia delivery, provides a special focus for provisioning QoS to the services that are delivered to the end user.

Since QoS is an integral part of delivering service, QoS control is a core part of IMS with the goal that the end-to-end level of QoS is maintained for a user across different access networks.

The rest of the design goals for IMS are in terms of creation, delivery and management of services. IMS is also supposed to enable fast creation and deployment of services using a common set of protocols. Delivery of service should allow service composition where the service architecture allows a combination of heterogeneous services under a single session, single authentication and charging model. This is similar to the way web-service architecture has evolved in delivering rich and flexible services from heterogeneous service components. Following the above design principles, the IMS architecture can be viewed as a set of subsystems where each subsystem provides a certain set of services/functionalities. The subsystems interact among themselves using standard signaling protocols over an IP network to create and deliver diverse services to end-users. However, it must be noted that the subsystems are logical entities and can reside on a single node as well.

18.3   IMS ADVANTAGES

The push towards adoption of the IMS architecture is driven by the advantages that IMS provides to users, network operators and service providers. We start with the users' perspective on how IMS can provide a better user experience.

18.3.1   End-user experience

Indeed, the end-user experience matters as at the end, if the users do not see any benefit, the technology cannot create any new value/revenue for the network operators. IMS completely changes the end user experience in terms of how the services are used.

IMS will allow reduction in the number of devices to access services since the same services can be provided over any underlying access technology. An IMS-compliant device can access all IMS-compliant applications in an identical way, regardless of the device used and access network connection. By being multimedia capable, an IMS device can handle voice, video and data anywhere, anytime and over any network. As noted above, the ubiquity in delivering services allows users seamlessly to move from one access network to another. Consequently, the same device can work as a home phone through a PSTN line when at home, or a cell phone using the cellular network when roaming or a WiFi phone when in the office or in the coverage of a WiFi hotspot. Furthermore, IMS allows multiple services to be blended over a single session, allowing the users to multitask on the same device. For example, one can be browsing for information while talking over VoIP to a friend.

18.3.2   Enterprise-user experience

Typically, an enterprise environment is required to deliver and support various enterprise-level services to its employees. Therefore, for an enterprise, delivering and supporting a wide range of services matters. In many cases, the applications are run by the enterprise inside the enterprise boundary with strict security provisions. IMS provides the enterprise users with simple ways of directly delivering enterprise applications and services, regardless of the method of access or location. For example, IMS can deliver internal company news in video to any device at any time, simply by registering the device as having access. IMS also allows various authentication and authorization features at session level, which is an important requirement in an enterprise setting.

18.3.3 Benefits for network operators

IMS provides several advantages to the network operators. With IMS, the network operators can directly export the transport level QoS to specific application and charge the cost of resource provisioning. With IMS, the network operators are not constrained to deliver their services to only those users who are connected to their own networks. For example, a cellular provider can easily extend their VoIP service to users through a fixed network when the user is located at home. In essence, it becomes possible for a network operator to increase the scope of his/her network coverage both in terms of data delivery and service and to increase the subscriber base.

18.3.4 Benefits for service providers

From the service provider's standpoint, the IMS infrastructure provides a platform for fast, inexpensive and flexible creation, deployment and delivery of rich services. As with traditional architecture, the service creation was integrated with the network and required development of separate modules related to authentication, billing, subscriber databases, etc. This resulted in a slow and expensive creation and deployment of services. With IMS, services can be combined and applications can make use of common services related to authentication, billing etc. Such an approach is similar to Service Oriented Architecture and can result in significant reduction in the capital and operational expenditure for a service provider. The result is rapid service deployment and an evolutionary approach in creating new services from an existing set of services. Such composition of services is supported in IMS due to the provision that services export standard interfaces and can talk directly to each other using standardized protocols.

Another big advantage of IMS for the service providers is in terms of the diverse class of multimedia services that can be delivered. IMS makes this possible by having applications specify resource requirements and meeting the resource requirements by having resource management hooks at the transport layer. In essence, IMS provides a hook from the application layer to the underlying network/transport layer in terms of QoS provisioning. This will allow support of VoIP and Video to users with a high quality of experience requirement.

18.4 IMS ARCHITECTURE ORGANIZATION

The functionalities of the IMS architecture can be understood in two ways: layered interaction model; and component/subsystem interaction model. The layered way of presenting IMS architecture helps to provide a high-level understanding of the functional boundaries. The component or subsystem interaction is important to understand the actual design of IMS, which is a set of subsystems with defined roles and interfaces.

First, we look at the layered architectural model of IMS as shown in Figure 18.1. The entire IMS can be divided into three basic layers as follows:

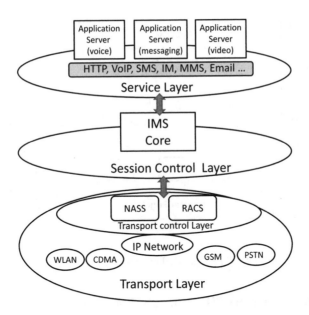

Figure 18.1 Layered architectural model of IMS.

- an *Application layer* consisting of Application Servers (AS) that host the IMS services and a Home Subscriber Server (HSS);

- a *Session Control layer* made up of several service subsystems which are part of the IMS core;

- a *Transport layer* consisting of two layers: (a) transport control layer, and (b) transport function layer. The transport control layer consists of two subsystems – the *Resource Admission Control Subsystem (RACS)* and the *Network Attachment Subsystem (NASS)*. The transport function layer consists of the actual network elements that carry the data.

The layering of the functionalities makes it easier to understand the operation of IMS. The lowest transport layer takes care of data transport and network-level resource allocation/management. The session control layer takes the responsibility of session-level authentication, authorization and admission. The session control layer interacts with the transport layer to request network-level resource allocation. The top layer consists of application-level services that utilize the underlying (session control and transport) layers to deliver services to the end user.

Each layer consists of several subsystems with defined roles as shown in Figure 18.2. There are many subsystems, among which the most important are the IMS core, NASS and the RACS. The figure shows the simplistic interaction model where the services interact with the IMS core. The IMS core contacts RACS for resource allocation. Both NASS and IMS cores use the subscriber profile database for making session-level decisions. Next we discuss the role of the important subsystems in more detail.

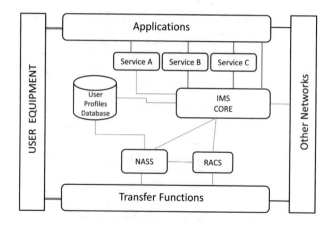

Figure 18.2   Core functional entities (subsystems) in IMS.

18.5   NETWORK ATTACHMENT SUBSYSTEM (NASS)

NASS [9] is a part of the Transport control layer. The role of NASS is to provide network-level identification and authentication. The role relates to providing registration and initialization of the user equipment so that the subscriber can access the network and the service over the network. The specific roles of NASS can be summarized as follows:

- dynamic provisioning of IP addresses and other terminal-configuration parameters;
- authentication at the IP layer prior to or during the address-allocation procedure.;
- authorization of network access based on user profiles;
- access network configuration based on user profiles;
- location management at the IP layer.

In order to provide the above roles, NASS is structured to have the following well-defined functional entities responsible for each role.

- *Network Access Configuration Function (NACF):* NACF is responsible for the IP address allocation to the User Equipment and can be implemented using the DHCP protocol.

- *Access Management Function (AMF):* AMF acts as an interface in translation and forwarding of user requests between the Access network and NACF.

- *Connectivity Session Location and Repository Function (CLF):* CLF is used to associate the user IP address to his location information and store other information about the user that includes profiles, preferences etc.

- *User Access Authorization Function (UAAF):* UAAF is used to authenticate users based on user profile stored and accessed through Profile Database Function (PDBF).

- *CNG Configuration Function (CNGCF):* CNGCF is used to configure the Customer Network Gateway (CNG) when necessary.

For more details on each of the functional entities, see [9].

18.6 RESOURCE ADMISSION CONTROL SUBSYSTEM (RACS)

RACS [8] has an important role in the overall IMS architecture where it provides two main functions: admission control, and network traffic control. RACS perform admission control based on (a) user profile stored at NASS, and (b) network operator-specific policies and resource availability. RACS exports the transport control services to the higher layers. Using these services, the higher layer can request and reserve transport resources. By using RACS, the upper layer can be agnostic about how the transport layer is used in allocating resources for the services delivered. Therefore, RACS is an important component enabling the IMS design goals.

RACS is interfaced with the higher layers using the Service-based Policy Decision Function (SPDF). Basically, SPDF provides a single point of contact for the high-level application. SPDF performs policy-based decisions for a given service-level request and sends appropriate resource-level requests to the Access-Resource Admission and Control Function (A-RACF). The A-RACF is responsible for managing resource reservation and admitting/rejecting the resource-level requests. Therefore, SPDF and A-RACF are the primary functional modules of RACS.

Finally, RACS also provides access to services provided by Border Gateway Functions such as NAT and hosted NAT traversal. More details about the role of RACS for QoS management is discussed later in the chapter.

18.7 IMS CORE SUBSYSTEM

The IMS core subsystem specified in TISPAN is the core component of the NGN architecture that extends the 3GPP IMS into wireline networks. IMS core provides control access to the SIP-based multimedia services. The IMS core consists of various control functional entities among which *call session control* is the most important one. These functional entities are described next.

18.7.1 Call session control

Call Session Control Function (CSCF) is mostly defined using the Session Initiation Protocol (SIP). SIP provides the signalling protocol. The job of the CSCF is to establish, monitor, support and release multimedia sessions and also manage the user's service interactions. The function of the CSCF can be broken down intro three important roles. These roles are defined in Proxy-CSCF (P-CSCF), Serving-CSCF (S-CSCF) and Interrogating-CSCF (I-CSCF) functional entities. The roles are described in detail next.

18.7.1.1 Proxy-CSCF   A Proxy-CSCF (P-CSCF) is a SIP proxy that is the first point of contact for the IMS terminal. It can be located either in the visited network (in full IMS networks) or in the home network (when the visited network is not yet IMS compliant). The P-CSCF is assigned to an IMS terminal during registration. This assignment does not change for the entire duration of the registration. The P-CSCF operates by sitting on the path of all signalling messages and by inspecting every message. Therefore, the P-CSCF can authenticate a particular user session, help in preventing attacks and can protect the privacy of the user (through proxy role). P-CSCF is thereby trusted by other nodes and a user once authenticated by P-CSCF does not go through further authentication by other nodes.

P-CSCF also provides several other functionalities that include:

- compression and decompression of SIP messages using SigComp;

- creating IPSec based secure connection with the user terminal;

- user charging.

18.7.1.2 Serving-CSCF A Serving-CSCF (S-CSCF) is the central node of the signalling plane. It is essentially a SIP server, and is always located in the home network. It uses Diameter interfaces to connect to the HSS for download and upload of user profiles from HSS. S-CSCF handles SIP registrations, which allows it to bind the user location (e.g. the IP address of the terminal) and the SIP address. Like P-CSCF, S-CSCF also sits in the path of all signaling messages. From this vantage point, the S-CSCF performs the following important functions: (a) it decides to which application server(s) to forward the SIP message in order to provide their services; (b) it provides routing services using ENUM lookups. There can be multiple S-CSCFs in the network for load distribution and high availability reasons. Therefore, assignment of a S-CSCF to a user is done by the HSS when it is queried by the Interrogating-CSCF (described next).

18.7.1.3 Interrogating-CSCF An I-CSCF (Interrogating-CSCF) is a SIP-based control functional entity which is located at the edge of an administrative domain. The IP address is published in the DNS of the domain. This allows a remote server to locate the I-CSCF and use it as a forwarding point for SIP packets to the domain in which the I-CSCF belongs. By querying the HSS, I-CSCF can obtain the corresponding S-CSCF to which to forward the incoming SIP message.

18.7.2 Other functional control entities

IMS core also hosts several other functional control entities as listed below:

- *Multimedia Resource Function Control (MRFC):* MRFC is used for controlling a Multimedia Resource Function Processor (MRFP) that essentially provides transcoding and content adaptation functionalities.

- *Breakout Gateway Control Function (BGCF):* BGCF selects the network in which PSTN breakout is to occur and where within the network the breakout is to occur. It selects the MGCF. This means that it is used for interworking with the circuit-switched domain.

- *Media Gateway Controller Function (MGCF):* MGCF is used to control a Media Gateway.

18.8   IMS QoS MANAGEMENT

Providing an end-to-end QoS guarantee to the service delivered using IMS is a key requirement for the IMS architecture. Since IMS is supposed to host a variety of services including voice, video, IM, etc, IMS must enable QoS control features that can be used for meeting the QoS requirements of these diverse applications. To that end, IMS has also categorized the multimedia applications into four general QoS classes:

- conversational (VoIP);

- streaming (push-to-talk);

- interactive (whiteboard collaboration);

- background (IM).

Definitely, the path towards adoption of IMS will result in fulfilling the long-awaited need for having an end-to-end QoS provisioning over an existing best-effort IP network.

18.9   QoS PROVISIONING APPROACH

There are two standard approaches for QoS control as also adopted by IMS: Guaranteed QoS and Relative QoS.

18.9.1   Guaranteed QoS

This type of QoS provisioning approach is required to support the conversational and streaming class of applications. Guaranteed QoS control ensures that service delivery receives absolute bounds on selected QoS parameters such as bandwidth, delay, jitter or packet loss. This type of QoS provisioning is achieved through two functions: admission control, and resource reservation. The admission control decision ensures that there exist available resources to support the given service. In the case of admission, resource reservation is required to ensure resources are not allocated to another service.

18.9.2   Relative QoS

This type of QoS provisioning approach is required to support the streaming, interactive and background class of applications. The relative QoS provisioning strategy ensures that traffic for the service is given some level of priority. This approach does not provide absolute bounds; instead, it provides an assurance of better service quality in case of congestion.

18.9.3   QoS control mechanism in IMS

The overall QoS control in IMS is executed at two layers: the session control layer, and the transport layer, as shown in Figure 18.3. We describe the role of each layer next.

18.9.3.1   Session control layer   When a session arrives associated with a service, the P-CSCF is the first entity to check the session and become aware of the session description

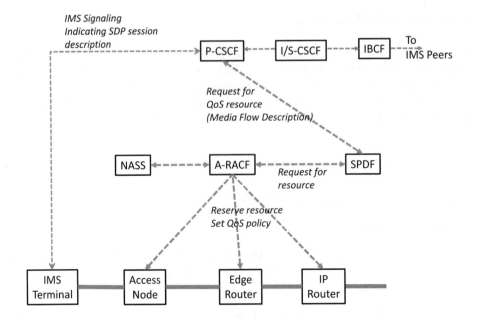

Figure 18.3   QoS control mechanisms in IMS.

protocol (SDP) context for the session. The SDP provides various information such as encoding rate and bandwidth requirement for the given multimedia session.

The IMS P-CSCF maps the information in the SDP to a media flow. The media flow has the information related to the maximum bandwidth requirement and QoS class (conversational, streaming, etc.). For reservation of QoS resources for an IMS session, the IMS P-CSCF indicates this list of media flows to the RACS.

RACS provides an interface to the IMS core through which the IMS can reserve network QoS resources at the transport layer.

18.9.3.2 *Transport layer*   Given the resource requirement from the IMS core, the role of the RAC is to perform resource admission control decision, and to reserve the required resource. Resource admission control refers to the decision to accept a given session based on the available resources. For admitted session through admission control, the RACS also enforces the decision by using rate control and policing.

The overall functionality of the RACS can be described in terms of the interaction between the subfunctional components of RACS. These functional components of RACS are: Access Resource Admission Control Function (A-RACF), Border Gateway Function (BGF) and Layer 2 Termination Point (L2TP).

A-RACF is the main component for the RACS for admission control. In order to make the decision, the A-RACF has a complete view of the available resource through interacting with the network management system. A-RACF decisions are dynamically enforced in the transport plane to make sure that service users do not exceed the granted resources. Enforcement is done through traffic policing by the BGF. BGF policies traffic at port level granularity for both downstream and upstream traffic.

18.9.4   Policy based QoS control

Policy based control is a check against a set of rules defined by the network operator for the service subscription of the subscriber. The purpose is to control the requests for QoS resources based on the policy rules set by the network operator. Policy control authorizes IMS media flows to run through the network only when they comply with the policy rules and can thus help prevent denial of service attacks or unauthorized users. Policy based control is generally the first stage of admitting the establishment of an IMS session.

18.10   SUMMARY

IMS is an all-IP-based architecture that will define the next generation networks in building a better service delivery model to end users. Alongside, IMS adoption will also lead the way for widespread adoption of VoIP. Due to a very modular and layered architecture consisting of interacting subsystems, it is easy to deploy IMS in an incremental fashion. IMS also provides a view of how the overall functionality of the network from service creation to resource provisioning can be integrated under a single architectural model. With steady growth in the space of multimedia applications with the recent explosion in video-based applications, IMS will be the inevitable solution for the network operators.

REFERENCES

1. 3GPP, Overview of 3GPP Release 5 – Summary of all Release 5 features, *3GPP - ETSI Mobile Competence Centre, Technical Report* (2003).

2. Camarillo, G. Introduction to TISPAN NGN, *Ericsson, Technical Report* (2005).

3. 3GPP, Technical specification group services and system aspects, IP Multimedia Subsystem (IMS), Stage 2, V5.15.0, TS 23.228 (2006).

4. Camarillo, G. and Garcia-Martin, M. The 3G IP Multimedia Subsystem (IMS): Merging the Internet and the cellular worlds, *John Wiley & Sons* (2006).

5. Poikselka, M., Niemi, A., Khartabil, H. and Mayer, G. The IMS: IP multimedia concepts and services, *John Wiley & Sons* (2006).

6. Zhuang, W., Gan, Y.S. and Chua, K.C. Policy based QoS architecture in the IP multimedia subsystem of UMTS, *IEEE Network*, 51–57 (May/June 2003).

7. Cuevas, A., Moreno, J.I., Vidales, P. and Einsiedler, H. The IMS service platform: A solution for next-generation network operators to be more than bit pipes, *IEEE Communications Magazine* August, 75-81 (2006).

8. TISPAN, ES 282 003: Resource and Admission Control Sub-system (RACS); functional architecture, *ETSI, Tech. Report* (2006).

9. TISPAN, ES 282 004: NGN functional architecture; Network Attachment Sub-System (NASS), *ETSI, Tech. Report* (2006).

INDEX